D0722859

Natural Systems for Waste Management and Treatment

Sherwood C. Reed

E. Joe Middlebrooks

Ronald W. Crites

McGraw-Hill Book Company

New York St. Louis San Francisco Auckland
Bogotá Hamburg London Madrid Mexico
Milan Montreal New Delhi Panama
Paris São Paulo Singapore
Sydney Tokyo Toronto

Library of Congress Cataloging-in-Publication Data

Reed, Sherwood C.
Natural systems for waste management and treatment.

Includes bibliographies and index.
1. Sewage disposal. 2. Sewage–Purification.
3. Factory and trade waste. 4. Refuse and refuse disposal. 5. Sewage sludge. I. Middlebrooks, E. Joe. II. Crites, Ronald W. III. Title.
TD645.R44 1988 628.4 87-3142
ISBN 0-07-051521-2

1234567890 DOCDOC 89210987

ISBN 0-07-051521-2

The editors for this book were Nadine M. Post and Jim Halston, the designer was Naomi Auerbach, and the production supervisor was Richard A. Ausburn. It was set in Century Schoolbook by Techna Type, Inc.
Printed and bound by R. R. Donnelley and Sons

Contents

Preface

This book is intended for the practicing engineer involved in the planning, design, construction, or operation of waste management facilities (both wastewater and sludges) for municipalities or industries.

The focus in this book is on waste management processes that depend to the maximum degree on natural components and to a minimal degree on mechanical elements. This utilization of natural systems can reduce the costs, process energy, and complexity of operation. These natural processes should be given priority consideration for planning new systems and for upgrading or retrofitting existing systems.

Some of the processes included in this book, such as pond systems, may be familiar to many engineers, but the text presents simplified, easy-to-use design procedures. The other, less familiar concepts can also provide very effective treatment for a significantly lower cost than mechanical treatment alternatives. Design criteria for some of the very new concepts cannot be found in any other text.

Each design chapter provides a complete description of the subject technology, data on performance expectations, and completely detailed design procedures with supporting examples. Chapter 2 presents the basic responses and interactions common to these natural biological systems. The treatment responses for toxic and hazardous materials are covered in this chapter and discussed it is as appropriate in the design chapters. Chapter 3 provides a rational procedure for planning and for process and site selection for the natural treatment systems. Combined metric and U.S. Customary System (USCS) units are used throughout the text, except in the design examples and tables, which are solved in metric units only. See the conversion tables in the Appendix for USCS equivalents for these items.

Sherwood C. Reed

Chapter

1

Natural Waste Treatment Systems—An Overview

The waste treatment systems described in this book are specifically designed to utilize natural responses to the maximum possible degree in obtaining the intended waste treatment or management goal. In most cases this will result in a system that costs less to build and to operate and requires less energy than mechanical treatment alternatives.

1.1 Natural Treatment Processes

All waste management processes depend on natural responses, such as gravity forces for sedimentation, or on natural components, such as biological organisms. However, in the typical case these natural components are supported by an often complex array of energy-intensive mechanical equipment. The term *natural system* as used in this text is intended to describe those processes that depend primarily on their natural components to achieve the intended purpose. A natural system might typically include pumps and piping for waste conveyance but would not depend on external energy sources exclusively to maintain the major treatment responses.

Background

Serious interest in natural methods for waste treatment reemerged in the United States following passage of the Clean Water Act of 1972 (Public Law 92-500). The major initial response was to assume that the "zero discharge" mandate of the law could be obtained via a combination of mechanical treatment units capable of *advanced wastewater treatment* (AWT). In theory, any specified level of water quality could be achieved with the proper sequence of mechanical operations. However, the energy requirements and high costs of this approach soon became apparent, and a search for alternatives commenced.

Land application of wastewater was the first *natural* technology to be rediscovered. In the nineteenth century it had been the only acceptable method for waste treatment, but it gradually slipped from use with the invention of modern devices. Studies and research quickly established that land treatment could realize all the goals of Public Law 92-500 while at the same time obtaining significant benefit from the reuse of the nutrients, other minerals, and organic matter in the wastes. Land treatment of wastewater became recognized and accepted by the engineering profession as a viable treatment concept during the decade following passage of Public Law 92-500, and it is now considered routinely in project planning and design.

Other natural concepts that had never dropped from use included lagoon systems and land application of sludges. Wastewater lagoons model the physical and biochemical interactions that occur in natural ponds, while land application of sludges models conventional farming practices with animal manures.

Aquaculture and wetland concepts are essentially new developments in the United States with respect to utilization of wastewaters and sludges. Much of the early developmental work focused on optimization of some other process component, with waste treatment considered incidentally. However, when specifically designed as a treatment process some concepts provide another cost-effective waste treatment option and are therefore included in this text. Several sludge management techniques, including conditioning, dewatering, disposal, and reuse methods, are also covered, since they also depend on natural components and processes.

Wastewater treatment concepts and performance expectations

Natural systems for effective wastewater treatment are available in three major categories: aquatic, terrestrial, and wetland concepts. All depend on natural physical and chemical responses as well as on the unique biological components in each process.

Aquatic treatment units. The design features and performance expectations for natural aquatic treatment units are summarized in Table 1.1. In all cases the major treatment responses are due to the biological components. Aquatic systems are further subdivided in the process design chapters to distinguish between lagoon or pond systems, which depend on microbial life and lower plants and animals, in contrast to aquaculture systems, which also utilize the higher plants and animals.

In most of the pond systems listed in Table 1.1 both performance and final water quality are dependent on the algae present in the system.[8] Algae are functionally beneficial, providing oxygen to support other biological responses, and the algae-carbonate reactions discussed in Chap. 4 are the basis for effective nitrogen removal in ponds. However, algae can be difficult to remove when stringent limits for suspended solids are required; alternatives to facultative ponds must then be considered. For this purpose controlled discharge systems were developed, in which, in essence, the treated wastewater is retained in the pond until the water quality in the pond and conditions in the receiving water are mutually compatible. The hyacinth ponds listed in Table 1.1 suppress algal growth in the pond because the plant leaves shade the surface and reduce penetration of sunlight. The other forms of vegetation and animal life used in aquaculture units are described in Chap. 5.

Wetland treatment units. Wetlands are defined as land in which the water table is at (or above) the ground surface long enough each year to maintain saturated soil conditions and the growth of related vegetation. The capability for wastewater renovation in wetlands has been verified in a number of studies in a variety of geographic settings. Wetlands used in this manner have included preexisting natural marshes, swamps, strands, bogs, peat lands, cypress domes, and systems especially constructed for wastewater treatment.

Table 1.2 summarizes the design features and expected performance for the three basic wetland categories. A major constraint on the use of many natural marshes is the fact that they are considered part of the receiving water by most regulatory authorities. As a result, the wastewater discharged to the wetland has to meet discharge standards prior to application to the wetland. In these cases the renovative potential of the wetland is not fully utilized.

Constructed wetland units avoid the special requirements on influent quality; they can also ensure much more reliable control over the hydraulic regime in the system and therefore perform more reliably than natural marshes. The rush or reed units listed in Table 1.2 are gravel- and sand-filled trenches or beds supporting a stand of appropriate vegetation. Their use for wastewater treatment is described in

TABLE 1.1 Design Features and Expected Performance for Aquatic Treatment Units[2,8,11]

Concepts	Treatment goals	Typical criteria*				Effluent characteristics, mg/L
		Climate needs	Detention time, days	Depth, m	Organic loading, kg/(ha · d)	
Oxidation pond	Secondary	Warm	10–40	1–1.5	40–120	BOD† 20–40 TSS‡ 80–140
Facultative pond	Secondary	None	25–180	1.5–2.5	22–67	BOD 30–40 TSS 40–100
Aerated pond partial mix	Secondary, polishing	None	7–20	2–6	50–200	BOD 30–40 TSS 30–60
Storage and controlled discharge ponds	Secondary, storage, polishing	None	100–200	3–5	—§	BOD 10–30 TSS 10–40
Hyacinth ponds	Secondary	Warm	30–50	< 1.5	< 30	BOD < 30 TSS < 30
Hyacinth ponds	AWT, with secondary input	Warm	> 6	< 1	< 50	BOD < 10 TSS < 10 TP¶ < 5 TN‖ < 5

* See Table A.1 in the appendix for conversion factors.
† BOD = biochemical oxygen demand.
‡ TSS = total suspended solids, concentration depends on algal content.
§ First cell in system designed as a facultative or aerated treatment unit.
¶ TP = total phosphorus.
‖ TN = total nitrogen (also get significant metals removal effected).

TABLE 1.2 Design Features and Expected Performance for Wetland Treatment Units[1,2]

Concepts	Treatment goals	Typical criteria*				Effluent characteristics, mg/L	
		Climate needs	Detention time, days	Depth, m	Hydraulic loading, m/(ha · day)		
Natural marshes	Polishing, AWT with secondary input	Warm	10	0.2–1	100	BOD†	5–10
						TSS†	5–15
						TN†	5–10
Constructed wetland	Secondary, or AWT	None	7	0.1–0.3	200	BOD	5–10
						TSS	5–15
						TN	5–10
Rush or reed beds	Secondary, or AWT	Warm	0.3	—‡	600	BOD	5–40
						TSS	5–20
						TN	5–20

* See Table A.1 in appendix for conversion factors.
† See Table 1.1 footnote for definition.
‡ Percolation and lateral flow in sand and gravel beds; see VSB in Chap. 6.

Chap. 6. A variation of this concept used for drying is described in Chap. 8.

Terrestrial treatment methods. Table 1.3 presents the typical design features and performance expectations for the three basic terrestrial concepts. All three are dependent on the physical, chemical, and biological reactions on and within the soil matrix. In addition, the *slow rate* (SR) and the *overland flow* (OF) methods require the presence of vegetation as a major treatment component. The SR process can utilize a wide range of vegetation, from trees to pastures to row crop vegetables. As described in Chap. 7, the OF process depends on perennial grasses to ensure a continuous vegetated cover. The hydraulic loading rates, with some exceptions, on rapid infiltration systems are typically too high to support beneficial vegetation. All three concepts can produce high-quality effluent. In the typical case the SR process can be designed to produce drinking water quality in the percolate.

Reuse of the treated water is possible with all three concepts. Recovery is easiest with OF since it is a surface system discharging to ditches at the toe of the treatment slopes. Most SR and rapid infiltration systems require underdrains or wells for water recovery.

Sludge management concepts. The freezing, composting, and reed bed concepts listed in Table 1.4 are intended to prepare the sludge for final disposal or reuse.

The freeze/thaw approach, described in Chap. 8, can easily increase sludge solids content to 35 percent or higher almost immediately upon thawing. Composting provides for further stabilization of the sludge and a significant reduction in pathogen content as well as reduction in moisture content. The major benefit of the reed bed approach is the possibility for multiple-year sludge applications and drying before removal is required. Solids concentrations acceptable for landfill disposal can be obtained readily.

Land application of sludge is designed to utilize the nutrient content in the sludge in agricultural, forest, and reclamation projects. Typically, the unit sludge loading is designed on the basis of the nutrient requirements for the vegetation of concern. The metal content of the sludge may then limit both the unit loading and the design application period for a particular site.

Costs and energy. Interest in natural concepts was originally based on the environmental ethic of recycle and reuse of resources wherever possible. Many of the concepts described in the previous sections do incorporate such potential. However, as more and more systems were built and operational experience accumulated, it was noticed that when

TABLE 1.3 **Design Features and Expected Performance for Terrestrial Treatment Units**[9]

Concepts	Treatment goals	Typical criteria: Climate needs	Vegetation	Area,* ha	Hydraulic loading, m/year	Effluent characteristics, mg/L	
Slow rate	Secondary, or AWT	Warmer seasons	Yes	23–280	0.5–6	BOD	< 2
						TSS	<1
						TN	< 3†
						TP	< 0.1
						FC	0‡
Rapid infiltration	Secondary, or AWT, or groundwater recharge	None	No	3–23	6–125	BOD	5
						FSS	2
						TN	10
						TP	< 1§
						FC	10
Overland flow	Secondary, nitrogen removal	Warmer seasons	Yes	6–40	3–20	BOD	10
						TSS	10¶
						TN	< 10
						TP	< 6

* For design flow of 3785 m^3/day.
† Nitrogen removal depends on type of crop and management.
‡ FC = fecal coliform, number per 100 mL.
§ Measured in immediate vicinity of basin; increased removal with longer travel distance.
¶ Total suspended solids depends in part on type of wastewater applied.

TABLE 1.4 Sludge Management with Natural Methods

Concept	Description	Limitations
Freezing	A method for conditioning and dewatering sludges in the winter months in cold climates. More effective and reliable than any of the available mechanical devices. Can use existing sand beds.	Must have freezing weather for long enough to completely freeze the design sludge layer.
Compost	A procedure to further stabilize and dewater sludges, with significant pathogen kill, so less restrictions on end use of final product.	Requires a bulking agent and mechanical equipment for mixing and sorting; winter operations can be difficult in cold climates
Reed beds	Narrow trenches or beds, with sand bottom and underdrained, planted with reeds. Vegetation assists water removal.	Best suited in warm to moderate climates. Annual harvest and disposal of vegetation is required.
Land application	Application of liquid or partially dried sludge on agricultural, forested, or reclamation land.	State and federal regulations limit the annual and the cumulative loading of some metals.

site conditions were favorable, these natural systems could usually be constructed and operated for less cost and with less energy than the more popular and more conventional mechanical technologies. Numerous comparisons have documented these cost and energy advantages,[7,13] which likely will remain and become even stronger over the long term. There were, for example, about 400 municipal land treatment systems using wastewater in the United States in the early 1970s. That number had grown to at least 1400 by the mid 1980s and is projected to pass 2000 by the year 2000. It is further estimated that a comparable number of private industrial and commercial systems exist. These process selection decisions have been and will continue to be made on the basis of costs and energy requirements.

1.2 Project Development

The development of a waste treatment project, either municipal or industrial, involves consideration of institutional and social issues in addition to the technical factors. These issues influence and can often control decisions during the planning and preliminary design stages. The current regulatory requirements at the federal, state, and local levels are particularly important. The engineer must determine these requirements at the earliest possible stage of project development to ensure that the concepts under consideration are institutionally feasible. References 3, 4, and 9 provide useful guidance on the institutional and social aspects of project development.

Table 1.5 provides summary guidance on the technical requirements for project development and indicates the chapter(s) in this book that provide the needed criteria. Detailed information on waste characterization and on the civil and mechanical engineering details of design are not unique to natural systems and are therefore not included in this text. References 5 and 6 are recommended for that purpose.

In all cases project development should consider the widest possible range of process alternatives during the early planning stages. A useful source for comparative, planning-level criteria on many mechanical treatment alternatives is the Computer-Assisted Design Procedure for the Design and Evaluation of Wastewater Treatment Systems (CAPDET). This computer program, available through the Environmental Protection Agency and elsewhere, permits the very rapid identification of potentially cost-effective processes. If site-specific criteria are not available, the program will generate estimates based on internal default values. It is suggested that the CAPDET program be used to identify several of the more cost-effective mechanical alternatives for comparison during the planning stage of the natural systems described in this text.

TABLE 1.5 Guide to Project Development

Task	Description	See Chapter
Characterize waste	Define the volume and the composition of the waste to be treated	*
Concept feasibility	Determine which if any of the natural systems are compatible for the particular waste and the site conditions and requirements	2, 3
Design limits	Determine the waste constituent that controls design	3
Process design	Pond systems	4
	Aquaculture units	5
	Wetland systems	6
	Terrestrial systems	7
	Sludge management	8
Civil and mechanical details	Collection network in the community, pump stations, transmission piping, etc	*

* Not covered in this text; see Refs. 5 and 6.

REFERENCES

1. Banks, L., and S. Davis: "Wastewater and Sludge Treatment by Rooted Aquatic Plants in Sand and Gravel Basins," *Proceedings of Workshop on Low Cost Wastewater Treatment,* Clemson University, Clemson, S.C., April 1983, pp. 205–218.
2. Bastian, R. K., and S. C. Reed (eds.): *Aquaculture Systems for Wastewater Treatment,* Environmental Protection Agency, EPA 430/9-80-006, September 1979.
3. Deese, P. L.: "Institutional Constraints and Public Acceptance Barriers to Utilization of Municipal Wastewater and Sludge for Land Reclamation and Biomass Production," *Utilization of Municipal Wastewater and Sludge for Land Reclamation and Biomass Production,* Environmental Protection Agency, EPA 430/9-81-013, July 1981.
4. Forster, D. L., and D. D. Southgate: "Institutions Constraining the Utilization of Municipal Wastewaters and Sludges on Land," *Proceedings of Workshop on Utilization of Municipal Wastewater and Sludge on Land,* University of California, Riverside, February 1983, pp. 29–45.
5. Metcalf & Eddy, Inc.: *Wastewater Engineering: Treatment, Disposal, Reuse,* 2d ed., McGraw-Hill, New York, 1979.
6. Metcalf & Eddy, Inc: *Wastewater Engineering: Collection and Pumping of Wastewater,* McGraw-Hill, New York, 1981.
7. Middlebrooks, E. J., C. H. Middlebrooks, and S. C. Reed: "Energy Requirements for Small Wastewater Treatment Systems," *J. Water Pollution Control Fed.,* vol. 53, no. 7, July 1981.
8. Middlebrooks, E. J., C. H. Middlebrooks, J. H. Reynolds, G. Z. Watters, S. C. Reed, and D. B. George: *Wastewater Stabilization Lagoon Design, Performance and Upgrading,* Macmillan, New York, 1982.
9. *Process Design Manual for Land Treatment of Municipal Wastewater,* Environmental Protection Agency, EPA 625/1-81-013, 1981.
10. *Process Design Manual for Land Treatment of Municipal Wastewater—Supplement on Rapid Infiltration and Overland Flow,* Environmental Protection Agency, EPA 625/1-81-013a, October 1984.
11. *Process Design Manual for Municipal Wastewater Stabilization Ponds,* Environmental Protection Agency, EPA 625/1-83-015, 1983.
12. Reed, S. C., R. Bastian, S. Black, and R. K. Khettry: "Wetlands for Wastewater Treatment in Cold Climates," *Proceedings Third American Water Works Assoc. Water Reuse Symp.,* American Water Works Assoc., Denver, Colo., August 1984.
13. Reed, S. C., R. W. Crites, R. E. Thomas, and A. B. Hais: *Cost of Land Treatment Systems,* Environmental Protection Agency, EPA 430/9-75-003, 1979.

Chapter

2

Planning, Feasibility Assessment, and Site Selection

It is important during the early planning stages of a waste management project to include as many alternatives as possible to ensure that the most cost-effective process is selected. The feasibility of the natural treatment processes described in this book depends significantly on site conditions, climate, and related factors. It is, however, not practical or economical to conduct extensive field investigations for every process at every potential site during planning and preliminary design. This chapter provides a sequential approach, in which the first step is to determine potential feasibility and the land area required for treatment and to identify possible sites. The second step is to evaluate these sites on the basis of technical and economic factors and select one or more for detailed investigation. The final step involves detailed field investigations, identification of the most cost-effective alternative, and development of the criteria needed for final design.

2.1 Concept Evaluation

A convenient starting point is to divide the many possible process concepts into *discharging* and *nondischarging* systems. The former systems, which typically have an outfall or other direct discharge to surface waters, would include treatment ponds, aquaculture, wetlands,

TABLE 2.1 Discharge Systems—Special Site Requirements

Concept	Requirement
Treatment ponds	Proximity to a surface water for discharge, impermeable soils or liner, no steep slopes, out of flood plain or diked, no bedrock or groundwater within excavation depth
Aquaculture	Same physical features as ponds; also must have suitable climate to support aquatic plants or other biological components
Constructed wetlands	Proximity to surface waters for discharge, impermeable soils or a liner, slopes 0–3%, not in flood plain, bedrock and groundwater below excavation depth
Overland flow	Relatively impermeable soils, clay and clay loams, slopes 0–15%, depth to groundwater and bedrock not critical but 0.5–1 m desirable, need access to surface water for discharge

and the overland flow (OF) land treatment concept. The second, nondischarging group includes the other land treatment concepts and the sludge treatment methods. Site topography, soils, geology, and groundwater conditions are important factors for the construction of discharging systems but are often critical components in the treatment process itself for nondischarging systems. Design features and performance expectations for both types of systems are given in Tables 1.1, 1.2, and 1.3, and other special characteristics and requirements are listed in Tables 2.1 and 2.2. It is presumed that any percolate from the nondischarging systems mingles with any groundwater that may be present and may then eventually emerge as subflow in adjacent surface waters. These systems are typically designed to satisfy regulatory water quality requirements in the percolate-groundwater mixture as it reaches the project boundary. Some of these concepts can also be designed as direct discharge systems if underdrains, recovery wells, or cutoff ditches are included as system components. The underdrained slow rate (SR) land treatment system at Muskegon, Michigan[4] is an example of this type, while the forested SR system in Clayton County, Georgia[7] depends on natural subsurface flow. This subflow does emerge in surface streams that are part of the community's drinking water supplies, but the land treatment system is not considered to be a discharging system as defined by the Environmental Protection Agency (EPA) and by the state of Georgia.

Resources required

A preliminary determination of process feasibility and the identification of potential sites is based on the analysis of maps and other existing

TABLE 2.2 Nondischarge Systems—Special Site Requirements

Concept	Requirement
	Wastewater Systems
Slow rate	Clay loams and sandy loams, > 0.15 to < 15 cm/h permeability preferred, bedrock and groundwater $>$ 1.5 m, slopes $< 20\%$, agricultural sites $< 12\%$.
Rapid infiltration	Sandy loams and sands, $>$ 5–50 cm/h permeability, bedrock and groundwater > 4.5 m, slopes $< 10\%$, sites where significant backfill required for construction should be avoided. Look for sites near surface waters or over non-drinking water aquifers.
	Sludge Systems
Land application	Generally the same as for agricultural or forested SR systems. See Chap. 8 for special requirements for toxic or hazardous sludges.
Composting, freezing, vermistabilization, or reed beds	Usually on wastewater treatment plant site, so special site investigation not required. All three require impermeable barrier to protect groundwater; freezing and reed beds also need underdrains for percolate.

information. The requirements in Tables 2.1 and 2.2, along with an estimate of the land area needed for each of the concepts, are used in this procedure. The community maps should show: topography; water bodies and streams; flood hazard zones; community layout and land use (e.g., residential, commercial, industrial, agricultural, forest, etc.); existing water supply and sewerage systems; anticipated areas of growth and expansion; and soil types within the community and adjacent areas. Sources for these maps include the U.S. Geologic Survey (USGS), the Soil Conservation Service (SCS), and state agencies, as well as local planning and zoning agencies.

Preliminary estimates of land area

The land area estimates derived in this section are used with the information in Tables 2.1 and 2.2 and the map study to determine if suitable sites exist for the process under consideration. These preliminary area estimates are very conservative; they are only intended for this preliminary evaluation and should not be used for final design.

Treatment ponds. The area estimate for treatment pond systems will depend on the effluent quality required [defined in terms of biochemical oxygen demand (BOD) and suspended solids (SS)] on the type of pond system proposed, and on the geographic location. A facultative pond in the southern United States, for example, will require less area than the same process in Canada. The equations given below are for total project area and include an allowance for dikes, roads, and unusable portions of the site.

Oxidation ponds. For a 1 m (3 ft) deep oxidation pond assumed to be in a warm climate, with 30-days detention, and organic loading of 90 kg/(ha · day) [80 lb/(acre · day)], the expected effluent quality is: BOD = 30 mg/L, SS > 30 mg/L.

$$A_{op} = kQ \tag{2.1}$$

where A_{op} = total project area, ha (acres)
Q = design flow, m³/day (gal/day)
$k = 3.2 \times 10^{-3}$ metric (3.0×10^{-5} USCS)

Facultative ponds in cold climates. Assuming more than 80 days detention in 1.5 m (5 ft) deep pond and organic loading of 16.8 kg/(ha · day) [15 lb/(acre · day)], the expected effluent quality is: BOD = 30 mg/L, SS > 30 mg/L.

$$A_{fc} = kQ \tag{2.2}$$

where A_{fc} = facultative pond site area, ha (acres)
Q = design flow, m³/day (gal/day)
$k = 1.68 \times 10^{-2}$ metric (1.60×10^{-4} USCS)

Facultative ponds in warm climates. Assuming more than 60 days detention, in a 1.5 m (5 ft) deep pond and organic loading of 56 kg/(ha · day) [50 lb/(acre · day)], the expected effluent quality is BOD = 30 mg/L, SS > 30 mg/L.

$$A_{fw} = kQ \tag{2.3}$$

where A_{fw} = facultative pond warm-climate site area, ha (acres)
Q = design flow, m³/day (gal/day)
$k = 5.1 \times 10^{-3}$ metric (4.8×10^{-5} USCS)

Controlled discharge ponds. Controlled discharge ponds are used in northern climates to avoid winter discharges and in warm areas to match effluent quality to acceptable stream flow conditions. The typical depth is 1.5 m (5 ft), maximum detention time 180 days, and expected effluent quality is: BOD < 30 mg/L, SS < 30 mg/L.

$$A_{cd} = kQ \tag{2.4}$$

where A_{cd} = controlled discharge pond, site area, ha (acres)
Q = design flow, m³/day (gal/day)
$k = 1.63 \times 10^{-2}$ metric (1.32×10^{-4} USCS)

Partial mix aerated ponds. The size of partial-mix aerated ponds will vary with climate. For this purpose assume more than 50-day detention, 2.5 m (8 ft) depth, and organic loading of 100 kg/(ha · day) [89 lb/(acre · day)]. Expected effluent quality is: BOD = 30 mg/L, SS > 30 mg/L.

$$A_{ap} = kQ \tag{2.5}$$

where A_{ap} = aerated pond, site area, ha (acres)
$k = 2.9 \times 10^{-3}$ (2.7×10^{-3} USCS)
Q = design flow, m^3/day (gal/day)

Hyacinth systems. Hyacinth systems can be designed for treatment of raw sewage or for any other treatment level, up to tertiary polishing of secondary effluent. As with other types of pond systems, the critical design parameter is organic loading. The degree of nutrient removal achieved with hyacinth systems is directly related to the frequency of harvest. Hyacinth systems are only practical in locations where the plant can survive naturally (see Fig. 2.1 for this range and Chap. 5 for detailed design criteria).

Secondary hyacinth ponds. Hyacinth ponds for secondary treatment are designed for a raw sewage input, detention time more than 50 days, depth 1.5 m (5 ft) or less, organic loading rate of 30 kg (ha · day) [27 lb/(acre · day)], and water temperature above 10°C. Expected effluent

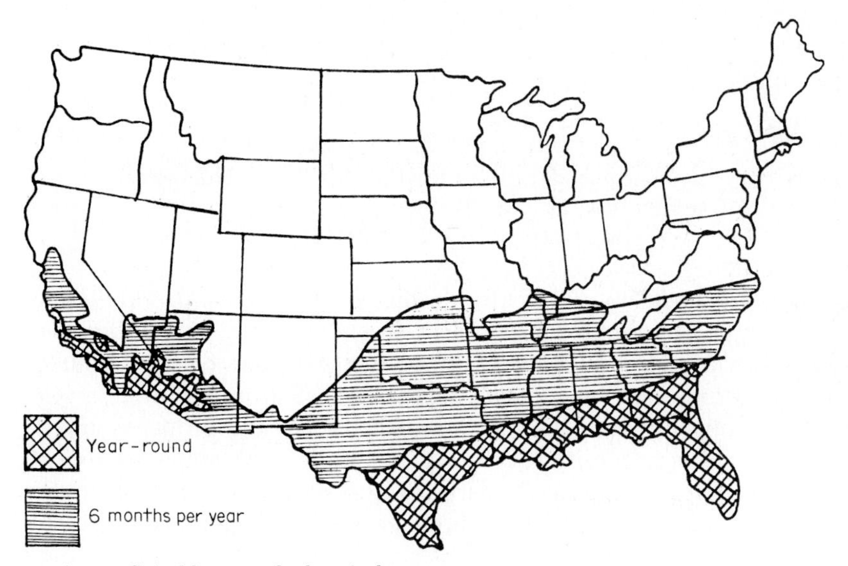

Figure 2.1 Suitable areas for hyacinth systems.

quality is: BOD < 30 mg/L, SS < 30 mg/L. A major function of the hyacinth plants is to suppress algae.

$$A_{hs} = kQ \tag{2.6}$$

where A_{hs} = hyacinth pond for secondary treatment, site area, ha (acre)
Q = design flow, m^3/day (gal/day)
k = 9.5×10^{-3} metric (8.9×10^{-5} USCS)

Advanced secondary hyacinth ponds. Advanced secondary hyacinth ponds are designed to provide better than secondary treatment, including some nutrient removal, with primary or equivalent input. Detention time more than 6 days, depth less than 1 m (3 ft), organic loading of 1000 kg/(ha · day) [900 lb/(acre · day)] supplemental aeration provided. Expected effluent quality is BOD < 10 mg/L, SS < 10 mg/L, with nitrogen removal dependent on frequency of harvest.

$$A_{has} = kQ \tag{2.7}$$

where A_{has} = hyacinth pond, advanced secondary treatment, site area, ha (acres)
Q = design flow, m^3/day (gal/day)
k = 9.5×10^{-4} metric (9.0×10^{-6} USCS)

Tertiary hyacinth ponds. Tertiary hyacinth ponds are designed to provide advanced treatment with secondary effluent input; other parameters are detention time more than 6 days, depth less than 1 m (3 ft), organic loading rate of 50 kg/(ha · day) [44.5 lb/(acre · day)], water temperature below 20°C, no supplemental aeration. Expected effluent quality: BOD < 10 mg/L, SS < 10 mg/L, total nitrogen < 5 mg/L, total phosphorus > 5 mg/L.

$$A_{ht} = kQ \tag{2.8}$$

where A_{ht} = hyacinth pond, tertiary treatment, site area, ha (acres)
Q = design flow, m^3/day (gal/day)
k = 7.1×10^{-4} metric (6.7×10^{-6} USCS)

Constructed wetlands. Typically designed for primary or equivalent quality input to produce better than secondary effluent quality, constructed wetlands can operate year-round even in coldest climates. Detention time is about 7 days, depth 0.1 m (0.3 m in winter in cold climates), and organic loading 23 kg/(ha · day) [20 lb/(acre · day)]. Expected effluent quality: BOD < 20 mg/L, SS < 20 mg/L, total N < 10 mg/L, total P > 5 mg/L. The area estimate given by Eq. 2.9 includes an allowance for an 8-day partial mix aerated pond as a pretreatment component.

$$A_{cw} = kQ \tag{2.9}$$

where A_{cw} = constructed wetland, site area, ha (acres)
k = 6.57×10^{-3} metric (6.0×10^{-5} USCS)
Q = design flow, m^3/day (gal/day)

Overland flow. The size of the overland flow (OF) project site will depend on the length of the operating season for this process. Figure 2.2 can be used to estimate the number of nonoperating days during which wastewater storage will be required. The design flow to the OF system is then calculated with Eq. 2.10.

$$Q_m = q_c + \frac{t_s q_c}{t_a} \tag{2.10}$$

where Q_m = average monthly design flow to land treatment site, m^3/month (gal/month)
q_c = average monthly flow in community, m^3/month (gal/month)
t_s = number of months storage required
t_a = number of months in operating season

The detention time on the OF slope is about 1 to 2 h, and depth of water on the slope is a few centimeters or less. The process design is not typically based on organic loading rates. Expected effluent quality is BOD = 10 mg/L, SS = 10 mg/L, total N $<$ 10 mg/L, total P $<$ 6 mg/L. The area estimate given by Eq. 2.11 includes an allowance for a 1-day aeration cell and for winter wastewater storage, if needed, as

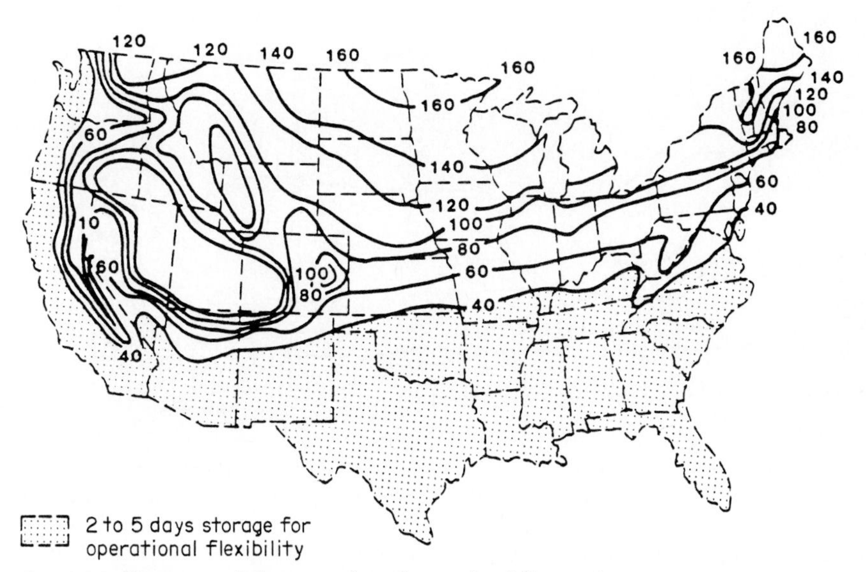

Figure 2.2 Recommended storage days for overland flow systems.

well as the actual treatment area, with an assumed 15 cm/week (6 in/week) hydraulic loading.

$$A_{of} = (3.9 \times 10^{-4})(Q_m + 0.05q_c t_s) \quad \text{(metric)}$$
$$A_{of} = (3.7 \times 10^{-6})(Q_m + 16.2q_c t_s) \quad \text{(USCS)} \tag{2.11}$$

where A_{of} = OF project area, ha (acres)
Q_m = average monthly design flow to land treatment site, m^3/month (gal/month)
q_c = average monthly flow in community, m^3/month (gal/month)
t_s = number of months storage required

Slow rate systems. Typically an SR system is a nondischarging system, and the size of the project site will depend on the operating season. Figure 2.3 can be used to estimate the number of operating months for locations in the United States.

The design flow to the SR system is then calculated with Eq. 2.12. Organic loading is not usually the critical design parameter. Either nitrogen or the hydraulic capacity of the soil will control for most municipal effluents (see Chap. 7); responses to industrial pollutants are considered in Chap. 3. The area estimate given by Eq. 2.12 includes an allowance for preapplication treatment in an aerated cell as well as a winter storage allowance and the actual land treatment area; a

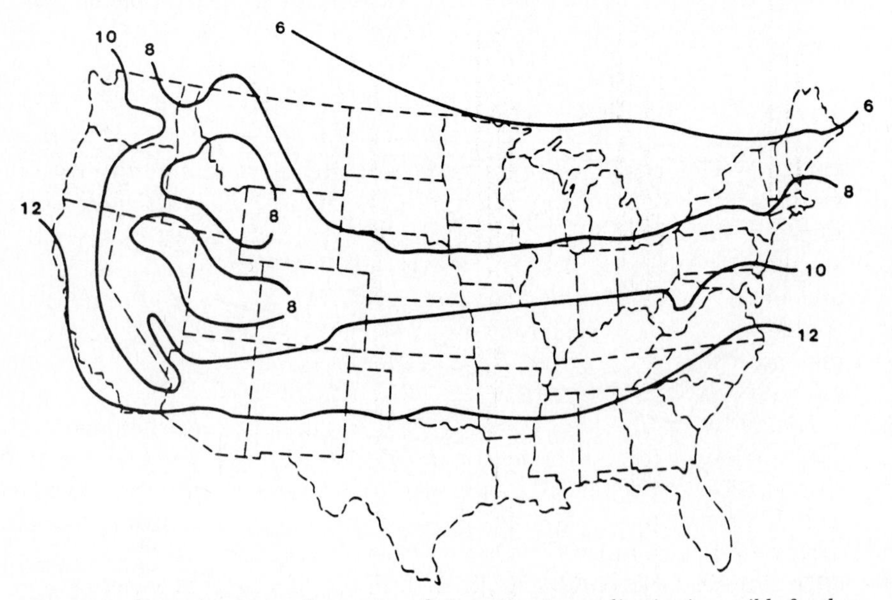

Figure 2.3 Approximate months per year that wastewater application is possible for slow rate systems.

hydraulic loading of 5 cm/week (2 in/week) was assumed. Expected effluent quality is BOD < 2 mg/L, SS < 1 mg/L, total N < 10 mg/L, total P < 0.1 mg/L.

$$A_{sr} = (6.0 \times 10^{-4})(Q_m + 0.03 q_c t_s) \quad \text{(metric)}$$
$$A_{sr} = (5.5 \times 10^{-6})(Q_m + 16.2 q_c t_s) \quad \text{(USCS)} \qquad (2.12)$$

where A_{sr} = SR system, total project area, ha (acres)
Q_m = average monthly design flow to land treatment site, m^3/month (gal/month)
q_c = average monthly flow in community, m^3/month (gal/month)
t_s = number of months storage required

Rapid infiltration systems. Typically a rapid infiltration (RI) system is a nondischarging system. Year-round operation is possible in all parts of the United States, so climate is not a factor in design; the design treatment area is usually controlled by the hydraulic capacity of the soils. As a result of the high hydraulic loadings, percolate nitrogen is likely to exceed 10 mg/L at times, and therefore sites should be selected where adverse impacts on drinking water aquifers will not occur. Expected percolate quality is BOD < 5 mg/L, SS < 2 mg/L, total nitrogen > 10 mg/L, and total phosphorus beneath basin < 1 mg/L. The area estimate given by Eq. 2.13 includes an allowance for preapplication treatment to the equivalent of primary quality.

$$A_{ri} = kQ_m \qquad (2.13)$$

where $k = 5.0 \times 10^{-5}$ metric (4.8×10^{-7} USCS)
A_{ri} = RI project area, ha (acres)
Q_m = average monthly design flow to land treatment site

Land area comparison. The land area required for a community wastewater flow of 4000 m^3/day (1.06 million gal/day) was estimated with Eqs. 2.1 to 2.13, for locations in a cold climate (5-month wastewater storage for SR and OF), in the mid-Atlantic states (3-month storage), and in the southern United States (no storage). The results are presented in Table 2.3. Allowances are made in the tabulated results for any preliminary treatment that might be required and for unused portions of the general site area.

Hyacinth systems are not considered outside the range shown in Fig. 2.1. Wetland systems are considered to be year-round operations for this purpose. Treatment responses in a wetland would proceed at a higher rate in warm climates, as described in Chap. 6, so a smaller area would be required as compared with a northern site. That difference is not critical at this stage of planning and process selection and is not included in Table 2.3.

TABLE 2.3 Land Area Estimates for 4000 m^3/day Systems

Treatment system	Area, ha		
	North	Mid-Atlantic	South
Pond systems			
Oxidation	NA*	NA*	12.8
Facultative	67.2	43.6	20.4
Controlled discharge	65.2	65.2	65.2
Partial-mix aerated	20.4	15.3	11.6
Hyacinth, secondary	NA	NA	38.0
Hyacinth, advanced secondary	NA	NA	4.0†
Hyacinth, tertiary	NA	NA	22.8‡
Constructed wetland	26.3	26.3	26.3
Slow rate	134.0	102.0	72.0
Overland flow	92.0	69.0	47.0
Rapid infiltration	6.0	6.0	6.0

* NA = not applicable.
† Includes allowance for primary treatment.
‡ Includes a 20-ha facultative pond.

Sludge systems. The land area required for composting, sludge freezing, vermistabilization, or reed bed dewatering is dependent on sludge quantity, moisture content, and local climate. It is necessary to use the procedures in Chap. 8 to determine the area required for each situation. A special site investigation may not be necessary since these sludge treatment facilities are usually located in the vicinity of the wastewater treatment facility and require only a minor portion of the total site area. A possible exception exists in the case of composting facilities for large quantities of sludge, where a remote site may be desirable to avoid residential complaints and to take advantage of lower land costs.

The area required for land application of sludges is dependent on sludge quantity and characteristics as well as on the type of operation intended. The loading rates in Table 2.4 can be used to estimate the land area required for each of the major land application options.

TABLE 2.4 Sludge Loadings for Preliminary Site Area Determination

Option*	Application schedule	Typical rate, metric tons/ha
Agricultural	Annual for 10 years	10
Forest	One time or at 5-year intervals for 20 years	45
Reclamation	One time	100
Type B	Annual	340

* See Chap. 8 for detailed description of options.

2.2 Site Identification

The information presented or developed in the previous sections is combined with the maps of the community area to determine if feasible sites for wastewater treatment or sludge disposal exist within a reasonable distance.

It is unlikely that a community or industry will have site conditions, within resonable proximity, for all the types of wastewater or sludge treatment facilities listed in Tables 2.1 and 2.2, and several will usually be dropped from consideration at an early stage. All the technically suitable sites should be located on the maps. In the next evaluation step local knowledge regarding land use commitments, costs, and the technical ranking procedure (described in the next section) are then considered to determine which process(es) and site(s) are technically feasible. A complex screening procedure is not usually required for the pond, aquaculture, and wetland concepts since the number of potential sites is usually limited. The critical factors in these cases are close proximity to the wastewater source and access to surface water for final discharge. The opposite is true for the concepts that involve land application of wastewater or sludge since a significant number of potential sites may exist. It will not be economical to conduct detailed site investigations on all potential sites, so a preliminary screening is justified.

Screening procedure

The screening procedure recommended by the EPA[13] utilizes rating factors to evaluate each potential site. Those sites with moderate to high scores are candidates for serious consideration and site investigation and testing. The conditions included in the general procedure include site grades, depth to groundwater, depth of soil, land use (present or future), and the pumping distance and elevation for the wastewater treatment concepts. The economical haul distance for sludge disposal and/or utilization concepts will depend on solids concentration and other local factors and must be determined on a case by case basis. Tables 2.5 and 2.6 are applicable for land application of wastewater.

Table 2.7 may be used for sludge concepts, and Tables 2.8 and 2.9 are for the special case of forested sites used for either sludge or wastewater treatment. The soil type is not included as a factor in Tables 2.5 to 2.7; it was included in Tables 2.1 and 2.2 and provides part of the basis for preliminary site identification, so it is not included again as a rating factor.

The relative importance of the various conditions in Tables 2.5 and 2.6 is reflected in the magnitude of the value assigned, so the largest value indicates the most important characteristic. The final category

TABLE 2.5 Physical Rating Factors for Land Application of Wastewater[16]

Condition	Concept		
	Slow rate	Overland flow	Rapid infiltration
Site grade, %			
0–5	8	8	8
5–10	6	5	4
10–15	4	2	1
15–20	Forest only 5	NS*	NS
20–30	Forest only 4	NS	NS
30–35	Forest only 2	NS	NS
> 35	Forest only 0	NS	NS
Soil depth, m†			
0.3–0.6	NS	0	NS
0.6–1.5	3	4	NS
1.5–3.0	8	7	4
> 3.0	9	7	8
Depth to groundwater, m			
< 1	0	5	NS
1–3	4	6	2
> 3	6	6	6
Soil permeability, cm/h (most restrictive layer)			
< 0.15	1	10	NS
0.15–0.50	3	8	NS
0.50–1.50	5	6	1
1.50–5.10	8	1	6
> 5.10	8	NS	9

* NS = not suitable.
† Soil depth to bedrock or impermeable barrier.

in Table 2.6 relates to the anticipated management of the land application site. It is possible under favorable conditions in rural areas to find farmers or forestry personnel who may be willing to accept wastewater or sludges for their nutrient value and who would prefer to continue to manage the site.

The ranking for a specific site is obtained by summing the individual values from Tables 2.5 and 2.6. The highest-ranking site will then be the most suitable. The suitability ranking can be evaluated with the following ranges:

Low suitability	< 18
Moderate suitability	18–34
High suitability	34–50

The restrictions on liquid (< 7 percent solids) sludges in Table 2.7 are to control runoff or erosion of surface-applied sludges. Injection of

TABLE 2.6 Land Use and Economic Factors for Land Application of Wastewater[16]

Condition	Rating value
Distance from wastewater source, km	
0–3	8
3–8	6
8–16	3
> 16	1
Elevation difference, m	
< 0	6
0–15	5
15–60	3
> 200	1
Land use, existing or planned	
Industrial	0
High density, residential or urban	0
Low density, residential or urban	1
Agricultural, or open space, for agricultural SR or OF	4
Forested, for forested sites	4
Forested, for agricultural SR or OF	1
Land cost and management	
No land cost, farmer or forest company management	5
Land purchased, farmer or forest company management	3
Land purchased, operated by industry or city	1

liquid sludges is acceptable on 6 to 12 percent slopes but is not recommended on higher grades without effective runoff control.

The values from Table 2.7 can be combined with the land use and land cost factors from Table 2.6 (if appropriate) to obtain an overall rating for a potential sludge application site. These combinations produce the following ranges:

	Agricultural	Reclamation	Type B
Low suitability	< 10	< 10	< 5
Moderate suitability	10–20	10–20	5–15
High suitability	20–35	20–35	15–25

The transport distance is a critical factor and must be included in the final ranking. The rating values for distance given in Table 2.6 can be used for agricultural sludge operations also. In general it is economical to transport liquid sludges (< 8 percent solids) about 16 km (10 mi) from the source. Dewatering is more cost-effective for greater distances.

TABLE 2.7 Physical Rating Factors for Land Application of Sludge[18]

Condition	Concept		
	Agricultural	Reclamation	Type B*
Site grade, %			
0–3	8	8	8
3–6	6	7	4
6–12 (no liquid sludge on ground surface)	4	6	NS†
12–15 (no liquid sludge)	3	5	NS
> 15 (no liquid sludge)	NS	4	NS
Soil depth, m‡			
< 0.6	NS	2	NS
0.6–1.2	3	5	2
> 1.2	8	8	8
Soil permeability, cm/h (most restrictive layer)			
< 0.08	1	3	5
0.08–0.24	3	4	5
0.24–0.8	5	5	5
0.8–2.4	3	4	0
> 2.4	1	0	NS
Depth to seasonal groundwater, m			
< 0.6	0	0	NS
0.6–1.2	4	4	2
> 1.2	6	6	6

* For surface treatment of industrial wastes. See Chap. 8 for details.
† NS = not suitable.
‡ Soil depth to bedrock or impermeable barrier.

Forested sites for either wastewater or sludge are presented as a separate category in Tables 2.8 and 2.9. In the earlier cases the type of vegetation to be used is a design decision to optimize treatment, and the appropriate vegetation is usually established during system construction. It is far more common for forested sites to depend on the preexisting vegetation on the site, so the type and status of that growth become important selection factors. The total rating combines values from Tables 2.8 and 2.9, and the final ranking, as with other concepts, must include transport distance. The values in Table 2.6 can be used for wastewater systems.

Climate

The regional climate has a direct influence on the sludge management options, as shown in Table 2.10. Climatic factors are not included in the rating procedure for wastewater systems since seasonal constraints on operations are already included as a factor in the land area deter-

TABLE 2.8 Rating Factors for Sludge or Wastewater in Forests, Surface Conditions[13]

Condition	Rating value
Dominant vegetation	
Pine	2
Hardwood or mixed	3
Vegetation age, years	
Pine > 30	3
20–30	3
< 20	4
Hardwood > 50	1
30–50	2
< 30	3
Mixed pine hardwood > 40	1
25–40	2
< 25	3
Slope, %	
> 35	0
0–1	2
2–6	4
7–35	6
Distance to surface waters, m	
15–30	1
30–60	2
> 60	3
Adjacent land use	
High density residential	1
Low density residential	2
Industrial	2
Undeveloped	3
RATING: 3–4 not suitable, 5–8 poor, 9–14 good, > 15 excellent	

minations. Seasonal constraints and the local climate are important factors in determining the design hydraulic loading rates and cycles for wastewater systems, as are the length of the operating season and storm water runoff conditions for all concepts. Table 2.11 lists the pertinent climatic data required for final design of both sludge and wastewater systems. At least a 10-year return period is recommended. References 8, 9, and 10 are useful sources of this information.

Flood hazard

The location of sludge or wastewater systems within a flood plain can either be an asset or a liability, depending on the approach used for planning and design. Flood-prone areas may be undesirable because of the variable drainage characteristics and the potential flood damage to the structural components of the system. On the other hand, flood plains and similar terrain may be the only deep soils in the area. If

TABLE 2.9 Rating Factors for Sludge or Wastewater in Forests, Subsurface Conditions[13]

Condition	Rating value
Depth to seasonal groundwater, m	
< 1	0
1–3	4
> 10	6
Depth to bedrock, m	
< 1.5	0
1.5–3	4
> 3	6
Type of bedrock	
Shale	2
Sandstone	4
Granite-gneiss	6
Rock outcrops, % of total surface	
> 33	0
10–33	2
1–10	4
None	6
SCS erosion classification	
Severely eroded	1
Eroded	2
Not eroded	3
SCS shrink-swell potential for the soil	
High	1
Low	2
Moderate	3
Soil cation exchange capacity, meq/100 g	
< 10	1
10–15	2
> 15	3
Hydraulic conductivity of soil, cm/h	
> 15	2
< 5	4
5–10	6
Surface infiltration rate, cm/h	
< 5	2
5–10	4
> 15	6
RATING: 5–10 not suitable, 15–25 poor, 25–30 good, 30–45 excellent	

permitted by the regulatory authorities, utilization of such sites for wastewater or sludge can be an integral part of a flood plain management plan. Offsite storage or wastewater or sludge can be a design feature to allow the site to flood as needed.

Maps of flood-prone areas have been produced by the U.S. Geological Survey (USGS) in many areas of the United States as part of the Uniform National Program for Managing Flood Losses. The maps are

TABLE 2.10 Climatic Influences on Land Application of Sludge[18]

Impact	Climatic region		
	Warm, arid	Warm, humid	Cold, humid
Operating time	Year-round	Seasonal	Seasonal
Operating cost	Lower	Higher	Higher
Sludge storage	Less	More	Most
Salt accumulation in soil	High	Low	Moderate
Leaching potential	Low	High	Moderate
Runoff potential	Low	High	High

based on the standard 7.5-min USGS topographic sheets, which identify the areas with a potential 1 in 100 chance of flooding in a given year by means of a black and white overprint. Other detailed flood information is usually available from local offices of the U.S. Army Corps of Engineers and of flood control districts. If the screening process identifies potential sites in flood-prone areas, the local authorities should be consulted to identify regulatory requirements prior to any detailed site investigation.

Water rights

Riparian water laws, primarily in states east of the Mississippi River, protect the rights of landowners along a watercourse to use of the water. The appropriative laws in the western states protect the rights of the prior users of the water. The adoption of any of the natural concepts for wastewater treatment can have a direct impact on water right concerns:

- Site drainage, both quantity and quality, may be affected.

TABLE 2.11 Climatic Data Required for Land Application Designs[13]

Condition	Required data	Type of analysis
Precipitation	Rain, snow, annual averages, maximums, minimums	Frequency
Storm events	Intensity, duration	Frequency
Temperature	Length of frost-free period	Frequency
Wind	Direction, velocity	Assess aerosol risk
Evapotranspiration	Annual and monthly averages	Annual distribution

- A nondischarging system or a new discharge location will affect the quantity of flow in the body of water previously used for discharge.
- Operational considerations for land treatment systems may alter the pattern and the quality of discharges to a water body.

In addition to surface waters in well-defined channels or basins, many states also regulate or control other superficial waters and groundwater. State and local discharge requirements for the appropriate situation should be determined prior to initiation of design. If the project has any potential for legal entanglement, a water rights attorney should be consulted.

2.3 Site Evaluation

The next phase of the site and system selection process involves field surveys to confirm map data and then field testing for verification and to provide the data needed for design. This preliminary procedure includes an estimate of capital and operation and maintenance costs so that the sites identified in previous steps can be evaluated for cost effectiveness. A concept and a site are then selected for final design based on these results. Each site evaluation must include the following information:

- Property ownership, physical dimensions of the site, current and future land use.
- Surface and groundwater conditions: location and depth of wells, surface waters, flooding and drainage problems, fluctuations in groundwater levels, quality and users of groundwater.
- Characterization of the soil profile to 1.5 m (5 ft) for SR and most sludge systems, to at least 3 m (10 ft) for RI and pond-type systems, both physical and chemical properties.
- Agricultural crops: cropping patterns, yields, fertilizers used, tillage and irrigation methods, end use of crop, vehicular access within site.
- Forest site: age and species of trees, commercial or recreational site, irrigation and fertilizer methods, vehicle access to and within site.
- Reclamation site: existing vegetation, historical causes for disturbance, previous reclamation efforts, need for regrading or terrain modification.

Investigation of RI sites requires special consideration of the topography and of soil type and uniformity. Extensive cut-and-fill or related earthmoving operations not only are expensive but can alter the necessary soil characteristics through compaction. Sites with significant and numerous changes in relief over a small area are not the best choice for RI. Any soil with a significant clay fraction (higher than 10 percent) would generally exclude RI construction if fill were required by the design. Extremely nonuniform soils over the site do not absolutely preclude development of an RI system, but they significantly increase the cost and complexity of site investigation.

Soils investigation

Table 2.12 presents a sequential approach to field testing to define the physical and chemical characteristics of the on-site soils. In addition to the on-site test pits and borings, the routine investigation should include examination of exposed soil profiles in road cuts, borrow pits, and plowed fields on or near the site.

Backhoe test pits to a 3 m (10 ft) depth where soil conditions permit are recommended in each of the major soil types on the site. Soil samples should be obtained from critical layers, particularly from the layer being considered as the infiltration surface for wastewater or application layer for sludge. These samples should be reserved for future testing. The walls of the test pit should be carefully examined to define the characteristics listed in Table 2.13; Refs. 11, 15, and 17 are useful sources for this purpose. The test pit should be left open long enough to determine if groundwater seepage occurs, and then the highest level attained should be recorded. Equally important is any indication of seasonally high groundwater, most typically demonstrated by mottling of the soils (see Ref. 15).

Soil borings should penetrate below the groundwater table if it is within 10 to 15 m (30 to 50 ft) of the surface. At least one boring should be located in every major soil type on the site. If generally uniform conditions prevail, this might mean one boring for every 1 to 2 ha (2 to 5 acres) for large-scale systems. For small systems (< 5 ha) three to five shallow borings spaced over the entire site should be considered (see Refs. 17 and 18).

All the parameters in Table 2.13 can be observed or estimated directly in the field by experienced personnel. This preliminary field identification serves to confirm or modify the published soils data obtained during the map survey phase. Laboratory tests with the reserved samples confirm the field identification and provide criteria for design.

Soil texture and structure. Soil texture and structure are particularly important when infiltration of water is a design factor. The textural classes and the general terms used in soil descriptions are listed in Table 2.14. *Soil structure* refers to the aggregation of soil particles into

TABLE 2.12 Sequence of Field Testing (Typical Order from Left to Right)[1]

Comments	Tests			
	Test pits →	Borings →	Infiltration tests* →	Soil chemistry†
Type of test	Backhoe pit, also inspect road cuts, etc.	Drilled or augered, also logs of local wells for soils data and water level	Basin method if possible	Also review SCS soils surveys
Data needed	Depth of profile, texture, structure, restricting layers	Depth to groundwater, depth to barrier	Infiltration rate	N, P, metals, etc. retention, soil and crop management
Then estimate	Need for hydraulic conductivity tests	Groundwater flow direction	Hydraulic capacity	Soil amendments, crop limitations
More tests for	Hydraulic conductivity, if needed	Horizontal conductivity, if needed	—	—
Also estimate	Loading rates	Groundwater mounding, need for drainage	—	Quality of any percolate
Number of tests	Minimum 3–5 per site, more for large sites, poor soil uniformity	Minimum 3 per site, more for RI than SR, more for poor soil uniformity	Minimum 2 per site, more for large sites or poor soil uniformity	Depends on type of site, soil uniformity, waste character

* Only required for land application of wastewater; some definition of subsurface permeability needed for pond and sludge systems.

† Typically only needed for land application of sludges or wastewaters.

TABLE 2.13 Soil Characteristics in Field Investigations[1]

Characteristic	Significance
Estimate % gravel, sand and fines	Influences permeability
Soil textural class	Influences permeability
Soil color	Indication of seasonal groundwater, soil minerals
Plasticity of fines	Permeability and influence on cut or fill earthwork
Stratigraphy and structure	Ability to move water vertically and laterally
Wetness and consistency	Drainage characteristics

clusters of larger particles called *peds*. Well-structured soils with large voids between peds will transmit water more rapidly than structureless soils of the same texture. Even fine-textured soils that are well structured can transmit large quantities of water. Earthmoving and related construction activity can alter or destroy the in situ soil structure and significantly change the natural permeability. Soil structure can be observed in the side walls of a test pit. References 9 and 12 are suggested for additional detail.

Soil chemistry. The chemical properties of a soil affect plant growth, control the removal of many waste constituents, and influence the hydraulic conductivity of the soil profile. Sodium, for example, can affect the permeability of soils by dispersing clay particles and thereby

TABLE 2.14 Soil Textural Classes and General Terminology Used in Soil Descriptions[15]

Common name	Texture	Class name	USCS* symbol
Sandy soils	Coarse	Sand Loamy sand	GW, GP, GM-d SW
	Moderately coarse	Sandy loam Fine sandy loam	SP, SM-d
Loamy soils	Medium	Very fine sandy loam, loam, silt loam, silt	MH, ML
	Moderately fine	Clay loam, sandy clay loam, silty clay loam	SC
Clayey soils	Fine	Sandy clay, silty clay, clay	CH, CL

* USCS = Unified Soil Classification System

changing the soil structure that initially allowed water movement. The problem is most severe in arid climates. Chapter 3 discusses these sodium relationships and the importance of soil pH and soil minerals in greater detail. The potential chemical interactions between the waste materials and the soil are particularly important for land application or containment of toxic and hazardous materials. Chapter 8 contains information on land application of toxic sludges.

If the proposed system involves land application of sludges or wastewater and that in turn depends on surface vegetation as a treatment component, then soil chemistry is a very important factor in the development and future maintenance of that vegetation. The following tests are suggested for each of the major soil types on the site:

- pH, cation exchange capacity (CEC), exchangeable sodium percentage (ESP) (in arid climates), background metals lead (Pb), zinc (Zn), copper (Cu), nickel (Ni), cadmium (Cd), electrical conductivity (EC) of soil solution
- Plant-available nitrogen (N), phosphorus (P), potassium (K), lime requirements for pH adjustment and maintenance

There are few standard test procedures for chemical analysis of soils; Refs. 2, 6, and 12 are suggested for this purpose. Table 2.15 can be used to interpret results of these chemical tests. The *cation exchange capacity* of a soil is a measure of the capacity of the negatively charged soil colloids to adsorb cations from the soil solution. This adsorption is not necessarily permanent since the cations can be replaced by others in the soil solution. These exchanges do not significantly alter the structure of the soil colloids. The percentage of the CEC that is occupied by a particular cation is termed the *percent saturation* for that cation. The sum of the exchangeable hydrogen, sodium, potassium, calcium, and magnesium, expressed as a percentage of the CEC, is called *percent base saturation*. There are optimum ranges for percent base saturation for various crop and soil combinations. It is important for calcium and magnesium to be the dominant cations rather than sodium or potassium. The cation distribution in the natural soil can be easily changed by the use of soil amendments such as lime or gypsum.

The nutrient status of the soil is important if vegetation is to become a component in the treatment system or if the soil system is to otherwise remove nitrogen and phosphorus. Potassium is also measured to ensure maintenance of a proper balance with the other nutrients. The N:P:K ratio for wastewaters and sludges is not always suitable for optimum crop growth, and there have been cases in which the addition of supplemental potassium was necessary. See Chap. 3 for a detailed discussion on nutrients.

TABLE 2.15 Interpretation of Soil Chemical Tests[16]

Parameter and test result	Interpretation
pH of saturated soil paste	
< 4.2	Too acid for most crops
5.2–5.5	Suitable for acid tolerant crops
5.5–8.4	Suitable for most crops
> 8.4	Too alkaline for most crops
CEC, meq/100 g	
1–10	Limited adsorption (sandy soils)
12–20	Moderate adsorption (silt loam)
> 20	High adsorption (clay and organic soils)
Exchangeable cations	Desirable range, as % of CEC
Sodium	< 5
Calcium	60–70
Potassium	5–10
ESP as a % of CEC	
< 5	Satisfactory
> 10	Reduced permeability in fine-textured soils
> 20	Reduced permeability in coarse soils
EC, mS/cm at 25% of saturation extract	
< 2	No salinity problems
2–4	Restricts growth of very sensitive crops
4–8	Restricts growth of many crops
8–16	Only salt-tolerant crops will grow
> 16	Very few salt-tolerant crops will grow

Infiltration and permeability

The ability of water to infiltrate the soil surface and then percolate vertically or laterally is a critical factor for most of the treatment concepts discussed in this book. On the one hand, excessive permeability can negate the design intentions for most ponds, wetlands, and OF systems. Insufficient permeability will limit the usefulness of SR and RI systems and result in undesirable waterlogged conditions for land application of sludges. The hydraulic properties of major concern are the ability of the soil surface to infiltrate water and the flow or retention of water within the soil profile. These factors are defined by the saturated permeability or hydraulic conductivity, the infiltration capacity, and the porosity, specific retention, and specific yield of the soil matrix.

Saturated permeability. A material is considered permeable if it contains interconnected pores, cracks, or other passageways through which water or gas can flow. *Hydraulic conductivity* (synonymous with *permeability* as used in this text) is a measure of the ability of liquids and gases to pass through soil. A preliminary estimate of permeability can be found in most SCS soil surveys. The final site and process selection and design should be based on appropriate field and laboratory tests to confirm the initial estimates. Table 2.16 lists the permeability classes as defined by the SCS.

Natural soils at the low end of the range are best suited for ponds, wetlands, OF systems, and treatment of industrial sludges that might have toxic components. Soils in the mid-range are well suited for SR systems and for land application of sludges. These soils can be rendered suitable for the former uses via amendments or special treatment. The soils at the upper end of the range are only suited for RI systems in their natural state but they can be made suitable for ponds, wetlands, or OF systems with construction of a proper liner.

The movement of water through soils can be defined with Darcy's equation:

$$q = \frac{Q}{A} = K\left(\frac{\Delta H}{\Delta L}\right) \tag{2.14}$$

where q = flux of water, the flow per unit cross-sectional area, cm/h (in/h)
Q = volume of flow per unit time, cm^3/h (in^3/h)
A = unit cross-sectional area, cm^2 (in^2)
K = permeability (hydraulic conductivity), cm/h (in/h)
H = total head, m (ft)
L = hydraulic path length, m (ft)
$\Delta H/\Delta L$ = hydraulic gradient

The total head H can be assumed to be the sum of the soil water pressure head (h) and the head due to gravity (Z), i.e., $H = h + Z$. When the flow path is essentially vertical, the hydraulic gradient is equal to unity, and the vertical permeability K_v is used in Eq. 2.14. When the flow path is essentially horizontal, the horizontal permeability K_h should be used. The permeability coefficient K is not a true constant but a rapidly changing function of water content. Even under saturated conditions the K value may change owing to swelling of clay particles and other factors, but for general engineering design purposes it can be considered a constant. The K_v value will not necessarily be equal to K_h for most soils. In general, the lateral K_h will be higher

TABLE 2.16 SCS Permeability Classes for Saturated Soil[1]

Soil permeability, cm/h	Class	Soil permeability, cm/h	Class
< 0.06	Very slow	5.10–15.20	Moderately rapid
0.15–0.51	Slow	15.20–50.0	Rapid
0.51–1.50	Moderately slow	> 50.0	Very rapid
1.50–5.10	Moderate		

TABLE 2.17 Measured Ratio of K_h to K_v[16]

K_h, m/day	K_h/K_v	Comments
42	2.0	Silty soil
75	2.0	—
56	4.4	—
100	7.0	Gravelly
72	20.0	Near terminal morain
72	10.0	Irregular succession of sand and gravel layers; from field measurements of K

since the interbedding of fine- and coarse-grained layers tends to restrict vertical flow. Typical values are given in Table 2.17.

Infiltration capacity. The *infiltration rate* of a soil is defined as the rate at which water enters the soil from the surface. When the soil profile is saturated and there is negligible ponding at the surface, the infiltration rate is equal to the effective saturated permeability or conductivity of the immediate soil profile.

Although the measured infiltration rate on a particular site may decrease with time owing to surface clogging, the subsurface vertical permeability at saturation will generally remain constant. As a result the short-term measurement of infiltration serves reasonably well as an estimate of the long-term saturated vertical permeability within the zone of influence for the test procedure being used.

Porosity. The ratio of voids to the total volume of the soil is referred to as the *soil porosity*. It is expressed either as a decimal fraction or as a percentage, as defined by Eq. 2.15.

$$n = \frac{V_t - V_s}{V_t} = \frac{V_v}{V_t} \tag{2.15}$$

where n = porosity, %
V_t = total unit volume of soil, m^3 (ft^3)
V_s = unit volume of soil fraction, m^3 (ft^3)
V_v = unit volume of voids, m^3 (ft^3)

Specific yield and specific retention. The porosity of a soil defines the maximum amount of water that a soil can contain when it is saturated. The *specific yield* is the portion of that water that will drain under the influence of gravity, and the *specific retention* is the portion that will remain as a film and in very small voids. The porosity, therefore, is the sum of the specific yield and specific retention. Figure 2.4 illustrates the relationship for typical California soils.

The specific yield is used in defining aquifer properties, particularly

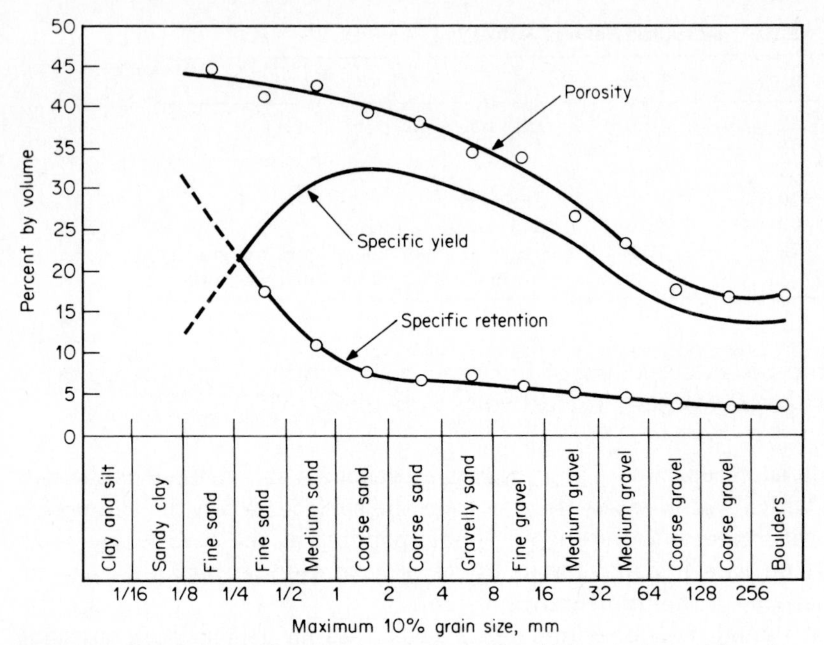

Figure 2.4 Porosity, specific retention, and specific yield variations with grain size, south coastal basin, California.

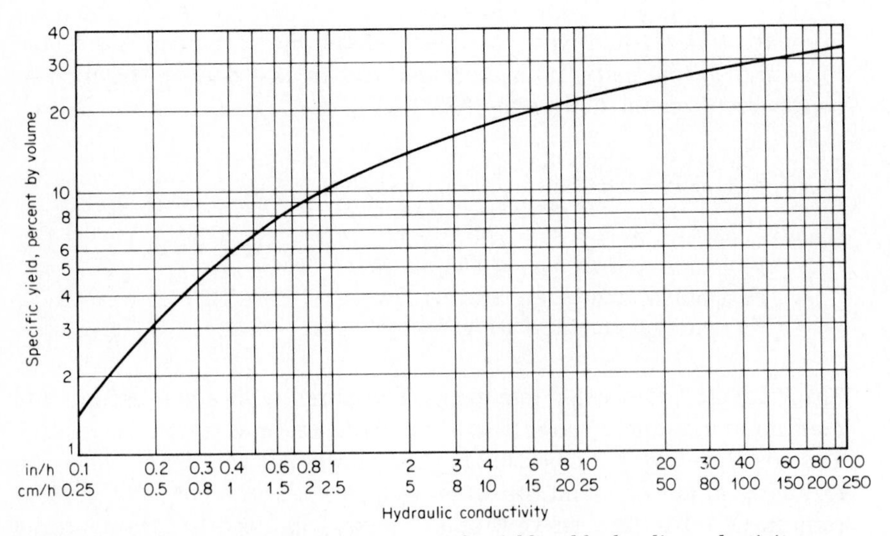

Figure 2.5 General relationship between specific yield and hydraulic conductivity.

in calculating groundwater mounding beneath ponds and wastewater application sites. For relatively coarse-textured soils and deep water tables it is acceptable to assume a constant value for the specific yield. Since the calculations are not especially sensitive to small changes in specific yield, it is usually satisfactory to estimate it from other properties, as shown on Figs. 2.4 and 2.5. Groundwater mound analysis can be more complicated for finer-textured soils because of capillarity effects in the soil as the water table moves higher (see Refs. 3 and 5 for details).

Test procedures and evaluation. In some cases it may be acceptable to utilize SCS estimates of soil permeability after confirming the actual presence of the specific soil on the site during a field investigation. This should be sufficient for pond and OF systems on soils with naturally low permeabilities. Systems in which water flow in the soil is a major design consideration will require field and possibly laboratory testing. Infiltration testing in the field is possible—and recommended—on the soil surface and, with smaller-scale devices, in test pits. It is necessary to utilize laboratory permeability tests on undisturbed soil samples from the test borings if deeper subsurface flow is a project concern. A variety of methods is available to measure infiltration rate or vertical saturated permeability K_v in the field. Some of the most common methods are compared in Table 2.18. The reliability of test results is a function of the test area and the zone of subsurface material influenced. This relationship is indirectly indicated in Table 2.18 by the volume of water required to conduct a single test. As indicated in Chap. 7, the increased confidence resulting from larger-scale field tests allows a reduction in the safety factor for the design of some land treatment systems.

TABLE 2.18 Comparison of Field Infiltration Testing Methods[1]

Technique	Water needs per test, L	Time per test, h	Equipment needed	Comments
Flooding basin	1900–7600	4–12	Backhoe or blade	See this chapter for details
Air entry permeameter (AEP)	10	0.5–1	AEP device	See this chapter for details
Cylinder infiltrometer	400–700	1–6	Standard device	See Ref. 16 for details
Sprinkler infiltrometer	1000–1200	1.5–3	Pump, pressure tank, sprinkler, collection cans	See Ref. 16 for details

Flooding basin test. A basin test area of at least 7 m^2 (75 ft^2) is suggested for all projects in which infiltration and percolation of water are design expectations. The area can be surrounded by a low earthen berm with an impermeable plastic cover, or aluminum flashing can be partially set into the soil in a circular configuration to define the test area. The use of a bentonite seal around the aluminum flashing perimeter is recommended to prevent leakage of water. Tensiometers at a depth of 15 cm (6 in) and 30 cm (12 in), near the center of the circle can be installed to define saturated conditions at these depths as the test progresses. The test basin should be flooded several times to ensure saturated conditions and to calibrate any instrumentation. The actual test run should be completed within 24 h of the preliminary trials and may require 3 to 8 h for coarse-textured soils.

The water level in the basin is observed and recorded with time, and the results are plotted as intake rate (in centimeters per hour) versus time. This intake rate will be relatively high initially and then drop off with time; the test must continue until this rate approaches a steady-state value, which can be taken as the limiting infiltration rate for the soil within the zone of influence of the test. A safety factor is then applied to that rate for system design, as described in Chap. 7.

Since it is the basic purpose of the test to define the hydraulic conductivity of the near-surface soil layers, the use of clean water (with about the same ionic composition as the expected wastewater) is acceptable in most cases. If, however, the wastewater is expected to have a high solids content that might clog the surface, a similar liquid should be used for the field test.

This basin test is most critical for the RI land treatment concept because large volumes of wastewater are applied to a relatively small surface area. Most RI systems, as described in Chap. 7, are operated on a cyclic pattern of flooding and drying to restore the infiltration capacity of the basin surface. If a particular project design calls for a continuously flooded *seepage pond* mode, the initial field tests should be continued for a long enough period to simulate this condition. If site conditions require the construction of the full-scale RI basins on backfilled material (not recommended), a test fill should be constructed on the site with the equipment intended for full-scale use, and the basin test described above should then be run in that material. The test fill should be as deep as required by the site design or 1.5 m (5 ft), whichever is less. The top of the fill area should be at least 5 m (16 ft) wide and 5 m (16 ft) long to permit the installation of a flooding basin test near the center.

One flooding basin infiltration test should be run on each of the major soil types on the site. For large continuous areas, one test for up to 10 ha (25 acres) is typically sufficient. The test should be performed on the soil layer that will become the final infiltration surface in the constructed system.

Air entry permeameter. The air entry permeameter was developed by the U.S. Department of Agriculture (USDA) to measure *point hydraulic conductivity* in the absence of a water table. The device is not commercially available, but specifications and fabrication details can be obtained from the U.S. Department of Agriculture, Water Conservation Laboratory, 4332 East Broadway, Phoenix, AZ 85040.

The unit defines conditions for a very small soil zone, but the small volume of water required and short time for a single test make it useful to verify site conditions between the larger-scale flooding basin tests. It can also be used in a test pit to define the in situ permeability with depth. The pit is dug with one end inclined to the surface, benches are cut about 1 m (3 ft) wide by hand, and the AEP is used on that surface.

Subsurface permeability and groundwater flow. The permeability of deeper soils is usually measured via laboratory tests on undisturbed soil samples obtained during the field boring program. Such data are usually only required for design of RI systems or to ensure that subsoils are adequate to contain undesirable leachates. In many situations it is desirable for the design of RI systems to determine the horizontal permeability of the subsurface layers. This can be accomplished with a field test called the *auger hole test,* which in essence consists of pumping a slug of water out of a borehole and then observing the time for the water level to recover via lateral flow. The U.S. Bureau of Reclamation has developed a standard procedure for this test, details of which can be found in Refs. 11 and 14.

Definition of the groundwater position and flow direction is essential for most of the treatment concepts discussed in this book. Overland flow and wetland systems have little concern with deep groundwater tables but might still be affected by near-surface seasonally high groundwater. Evidence of seasonal groundwater may be observed in the test pits, and water levels should be observed in any borings and in any existing wells on site or on adjacent properties. These data can provide information on the general hydraulic gradient and flow direction for the area. These data are also necessary if groundwater mounding or underdrainage, as described in Chap. 3, are project concerns.

Buffer zones

Prior to the site investigation, the state and local requirements for buffer zones or setback distances should be determined to be sure that adequate area exists on site or can be obtained. Most requirements for buffer zones or separation distances are based on aesthetics and the

TABLE 2.19 Setback Recommendations for Sludge Systems[18]

Setback distance, m	Suitable activities
15–60	Sludge injection, and only near: remote single dwellings, small ponds, 10-year high water mark for streams, roads. No surface applications.
90–460	Injection or surface application near all the above, plus springs and water supply wells; injection only near high-density residential developments.
> 460	Injection or surface application at all the above.

avoidance of odor complaints. The potential for aerosol transmission of pathogens is a concern to some for the operation of land application of wastewater and for some types of sludge composting operations (see Chap. 8 for a discussion of the latter). A number of aerosol studies have been conducted at both conventional and land treatment facilities with no evidence of significant risk to adjacent populations. Extensive buffer zones for aerosol containment are not recommended. If the system will utilize sprinklers, a buffer zone to catch the sprinkler droplets on windy days should be considered. A strip 10 to 15 m (32 to 49 ft) wide planted with conifers should suffice. Odor potential is the major concern for pond systems of the facultative type since the seasonal overturn may bring anaerobic materials to the liquid surface for a short period each spring and fall. A typical requirement in these cases is to locate such ponds at least 0.4 km (0.25 mi) from habitations. Mosquito control for wetland systems may require similar separation distances unless positive control measures are planned for the system. Recommended setback distances for sludge systems are listed in Table 2.19.

2.4 Site and Process Selection

The evaluation procedure discussed to this point has been to identify potential sites for a particular treatment alternative and then conduct field investigations to obtain data for the feasibility determination. The evaluation of the field data will indicate whether the site requirements listed in Tables 2.1 and 2.2 exist or not. If site conditions are favorable, it can be concluded that the site is apparently feasible for the intended treatment.

If only one site and the related treatment concept result from this screening process, then the focus can shift to final design and possibly additional detailed field tests to support that design. If more than one site for a particular treatment and/or more than one treatment remain technically viable after the screening process, it will be necessary to do a preliminary cost analysis to identify the most cost-effective alternative.

The criteria in Chaps. 4 through 8 should be used for preliminary design of the treatment alternative in question. Equations 2.1 to 2.13 in this chapter should not be used for this purpose, as they are only intended for a very preliminary estimate of the total amount of land that would be required for a particular treatment. The preliminary design should then be used as the basis for a preliminary cost estimate (capital and operating and maintenance), which should include land costs as well as pumping or transport costs to move the wastes from their source to the site. A comparison of these cost data will indicate the most cost-effective alternative. In many cases the final selection will also be influenced by the social and institutional acceptability of the proposed site and type of facility to be developed on it.

REFERENCES

1. Asano, T., and G. S. Pettygrove (eds.): *Irrigation with Reclaimed Municipal Wastewater—A Guidance Manual,* Water Resources Board, State of California, Sacramento, Calif., July 1984.
2. Black, A. (ed.): *Methods of Soil Analysis,* pt. 2: *Chemical and Microbiological Properties,* Agronomy 9, American Society of Agronomy, Madison, Wis., 1965.
3. Childs, E. C.: *An Introduction to the Physical Basis of Soil Water Phenomena,* Wiley, London, 1969.
4. Demirjiian, Y. A. et al.: Muskegon County Wastewater Management System—Progress Report, 1968–1975, Environmental Protection Agency Region 5, EPA 905/2-80-004, Chicago, Ill., February 1980.
5. Duke, H. R.: "Capillary Properties of Soils—Influence Upon Scientific Yields," *Transcripts Am. Soc. Agr. Engrs.,* 15:688–691, 1972.
6. Jackson, M. L.: *Soil Chemical Properties,* Prentice-Hall, Englewood Cliffs, N.J., 1958.
7. McKim, H. L. et al.: *Wastewater Applications in Forest Ecosystems,* U.S. Cold Regions Research and Engineering Laboratory, CRREL Report 82-19, Hanover, N.H., 1982.
8. National Oceanic and Atmospheric Administration: *The Climatic Summary of the United States,* Rockville, Md.
9. National Oceanic and Atmospheric Administration: *The Monthly Summary of Climatic Data,* Rockville, Md.
10. National Oceanic and Atmospheric Administration: *Local Climatological Data,* Rockville, Md.
11. Reed, S. C., and R. W. Crites: *Handbook of Land Treatment Systems for Industrial and Municipal Wastes,* Noyes Publications, Park Ridge, N.J., 1984.
12. Richards, L. A. (ed.): *Diagnosis and Improvement of Saline and Alkali Soils,* Agricultural Handbook No. 60, U.S. Department of Agriculture, Washington, D.C., 1954.
13. Taylor, G. L.: "A Preliminary Site Evaluation Method for Treatment of Municipal Wastewater by Spray Irrigation of Forest Land," *Proc. Conf. Appl. Research and Practice Municipal and Industrial Waste,* Madison, Wis., Sept. 1980.
14. U.S. Department of the Interior, Bureau of Reclamation: *Drainage Manual,* U.S. Government Printing Office, Washington, D.C., 1978.
15. U.S. Environmental Protection Agency: *Design Manual—Onsite Wastewater Treatment and Disposal Systems,* EPA 625/1-80-012, Water Engineering Research Laboratory, Cincinnati, Ohio, October 1980.

16. U.S. Environmental Protection Agency: *Process Design Manual—Land Treatment of Municipal Wastewater,* EPA 625/1-81-013, Center for Environmental Research Information, Cincinnati, Ohio, Oct. 1981.
17. U.S. Environmental Protection Agency: *Process Design Manual Supplement on Rapid Infiltration and Overland Flow,* EPA 625/1-81-013a, Center for Environmental Research Information, Cincinnati, Ohio, Oct. 1984.
18. U.S. Environmental Protection Agency: *Process Design Manual Land Application of Municipal Sludge,* EPA 625/1-83-016, Center for Environmental Research Information, Cincinnati, Ohio, Oct. 1983.

Chapter

3

Basic Process Responses and Interactions

It is the purpose of this chapter to describe the basic responses and interactions between the waste constituents and the process components of natural treatment systems. Many of these responses are common to more than one of the treatment concepts and are grouped in this chapter. If a waste constituent is the limiting factor for design, it is also discussed in detail in the appropriate process design chapters that follow.

Water is the major constituent in all the wastes of concern in this book, since even a "dried" sludge can contain more than 50 percent water. The presence of water is a volumetric concern for all treatment methods. It has a far greater significance for many of the natural treatment concepts, since the flow path and flow rate can control the successful performance of the system.

The other waste constituents of major concern include the simple carbonaceous organics (dissolved and suspended), toxic and hazardous organics, pathogens, trace metals, nutrients (nitrogen, phosphorus, potassium), and other micronutrients. The natural system components that provide the critical reactions and responses include bacteria, protozoa, algae, vegetation (aquatic and terrestrial), and the soil. The responses involved include a range of physical, chemical, and biological reactions.

3.1 Water Management

The major concerns include the potential for travel of contaminants with the groundwater, the risk of leakage from ponds and other aquatic systems, the potential for groundwater mounding beneath a land treatment system, the need for drainage, and the maintenance of design flow conditions in ponds, wetlands, and other aquatic systems.

Fundamental relationships

Chapter 2 introduced some of the hydraulic parameters (permeability, etc.) important to natural systems and discussed methods for their determination in the field or laboratory. It is necessary to provide further details and definitions prior to undertaking any flow analysis.

Permeability. The results from the field and laboratory test program described in Chap. 2 may vary with respect to both depth and areal extent even if the same basic soil type is known to exist over much of the site. The soil layer with the most restrictive permeability is taken as the design basis for those systems that depend on infiltration and percolation of water as a process requirement. In other cases in which there is considerable scatter to the data it is necessary to determine a *mean* permeability for design.

If the soil is uniform, then the vertical permeability K_v should be constant with depth and area, and any differences in test results should be due to variations in the test procedure. In this case the K_v can be considered to be the arithmetic mean, as defined by Eq. 3.1:

$$K_{am} = \frac{K_1 + K_2 + K_3 + \ldots + K_n}{n} \tag{3.1}$$

where K_{am} is the arithmetic mean vertical permeability and K_1 to K_n are individual test results.

In the case in which the soil profile consists of a layered series of uniform soils, each with a distinct K_v generally decreasing with depth, the average value can be represented as the harmonic mean:

$$K_{hm} = \frac{D}{\dfrac{d_1}{K_1} + \dfrac{d_2}{K_2} + \dfrac{d_n}{K_n}} \tag{3.2}$$

where D = soil profile depth
d_n = depth of nth layer
K_{hm} = harmonic mean permeability

If no pattern or preference is indicated by a statistical analysis, then a random distribution of the K_v values for a layer must be assumed, and the geometric mean provides the most conservative estimate of the true K_v.

$$K_{gm} = [(K_1)(K_2)(K_3) \ldots (K_n)]^{1/n} \tag{3.3}$$

where K_{gm} is the geometric mean permeability and K_1 to K_n are individual test results.

Equation 3.1 or 3.3 can also be used with appropriate data to determine the lateral permeability K_h. Typical values for the ratio K_h/K_v have been presented in Table 2.17.

Groundwater flow velocity. The flow velocity in a groundwater system can be obtained by combining Darcy's law, the basic velocity equation from hydraulics, with the soil porosity because flow can only occur in the pore spaces in the soil.

$$V = \frac{K_h \, \Delta H}{n \, \Delta L} \tag{3.4}$$

where V = groundwater flow velocity, m/day
K_h = horizontal saturated permeability, m/day
$\Delta H/\Delta L$ = hydraulic gradient, m/m
n = porosity, (as a decimal fraction; see Fig. 2.4 for typical values)

Equation 3.4 can be used to determine vertical flow velocity also. In this case the hydraulic gradient is equal to unity, and K_v should be used in the equation.

Aquifer transmissivity. The transmissivity of an aquifer is the product of the permeability of the material and the saturated thickness of the aquifer. In effect it represents the ability of a unit width of the aquifer to transmit water. The volume of water moving through this unit width can be calculated with Eq. 3.5.

$$q = K_h b w \frac{\Delta H}{\Delta L} \tag{3.5}$$

where q = volume of water moving through aquifer, m^3/day
b = depth of saturated thickness of aquifer, m
w = width of aquifer (for unit width $w = 1$), m
K_h = horizontal saturated permeability, m/day
$\Delta H/\Delta L$ = hydraulic gradient, m/m

In many situations well pumping tests are used to define aquifer properties. The transmissivity of the aquifer can be estimated by using pumping rate and drawdown data from well tests; Refs. 6 and 32 provide details.

Dispersion. The dispersion of contaminants in the groundwater is due to a combination of molecular diffusion and hydrodynamic mixing. The net result is for the concentration of the material to be less but the zone of contact to be greater at down-gradient locations. Dispersion will occur in a longitudinal direction (D_x) and transverse to the flow path (D_y). Dye studies in homogeneous and isotropic granular media have indicated that dispersion occurs in the shape of a cone about 6° wide from the application point.[10] Stratification and other areal differences in the field will typically result in much greater lateral and longitudinal dispersion. For example, the divergence of the cone could be 20° or more in fractured rock.[6] The dispersion coefficient is related to the seepage velocity as described with Eq. 3.6.

$$D = av \tag{3.6}$$

where D = dispersion coefficient, D_x longitudinal, D_y transverse, m^2/day
a = dispersivity, a_x longitudinal, a_y transverse, m
$v = V/n$ = seepage velocity of groundwater system, m/day
V = Darcy's velocity (from Eq. 3.5)
n = porosity (see Fig. 2.4 for typical values)

The dispersivity is difficult to measure in the field or to determine in the laboratory. It is usually measured in the field by addition of a tracer at the source, followed by observation of the concentration at surrounding monitoring wells. An average dispersion coefficient value of 10 m^2/day resulted from field experiments at the Ft. Devens, Massachusetts rapid infiltration (RI) system,[3] but predicted levels of contaminant transport changed very little when the assumed dispersivity was increased by 100 percent or more. Many of the values reported in the literature are site-specific, "fitted" values and cannot be reliably used for projects elsewhere.

Retardation. The hydrodynamic dispersion discussed in the previous section will affect all contaminant concentrations equally. However, adsorption, precipitation, and chemical reactions with other groundwater constituents will retard the rate of advance of the affected contaminants. This is described by the retardation factor R_d, the value of which can range from 1 to 50 for organics often encountered at field sites. The lowest values are for conservative substances such as chlorides, which are not removed in the groundwater system. Chlorides will move with the same velocity as the adjacent water in the system and any change in observed chloride concentration is due to dispersion only, not retardation. Retardation is a function of soil and groundwater characteristics and not necessarily a constant for all locations. The R_d for some metals might be close to 1 if the aquifer is flowing through

TABLE 3.1 Retardation Factors for Chloride and Selected Organic Compounds

Material	Retardation factor R_d	Material	Retardation factor R_d
Chloride	1	Dichlorobenzene	14
Chloroform	3	Styrene	31
Tetrachloroethylene	9	Chlorobenzene	35
Toluene	3		

clean, sandy soils with a low pH but close to 50 for clayey soils. The R_d for organic compounds is dependent on sorption of the compounds on soil organic matter plus volatilization and biodegradation. The sorptive reactions depend on the quantity of organic matter in the soil and on its solubility in the groundwater. Insoluble compounds such as DDT, benzo[a]pyrenes, and some polychlorinated biphenyls (PCBs) are effectively removed by most soils. More soluble compounds such as chloroform, benzene, and toluene are removed less efficiently by even highly organic soils. Because volatilization and biodegradation are not necessarily dependent on soil type, the removal of organic compounds via these methods tends to be more uniform from site to site. Table 3.1 presents retardation factors for a number of organic compounds as estimated from several literature sources.[3,10,27]

Movement of pollutants

The movement or migration of pollutants with the groundwater is controlled by the factors discussed in the previous section. This could be a concern for ponds and other aquatic systems as well as for the slow rate (SR) and RI land treatment methods. Figure 3.1 illustrates the subsurface zone of influence for an RI basin system or a treatment pond in which significant seepage is allowed.

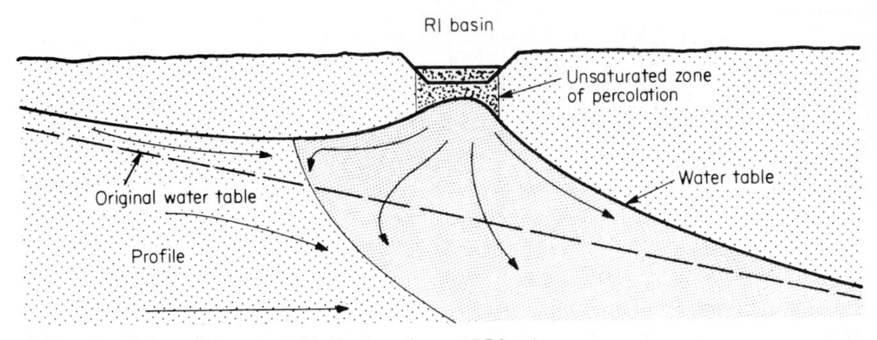

Figure 3.1 Subsurface zone of influence for an RI basin.

It is frequently necessary to determine the concentration of a pollutant in the groundwater plume at a selected distance down gradient from the source. Alternatively, it may be desired to determine the distance at which a given concentration will be present at a given time or the time at which a given concentration will reach a particular point. Figure 3.2 is a nomograph, which can be used to estimate these factors on the centerline of the down-gradient plume. The dispersion and retardation factors discussed above are included in the solution. Data required for use of the nomograph include:

- Aquifer thickness z, m
- Porosity n, as a decimal fraction
- Seepage velocity v, m/day
- Dispersivity factors a_x, a_y, m
- Retardation factor R_d for the contaminant of concern
- Volumetric water flow rate Q, m^3/day

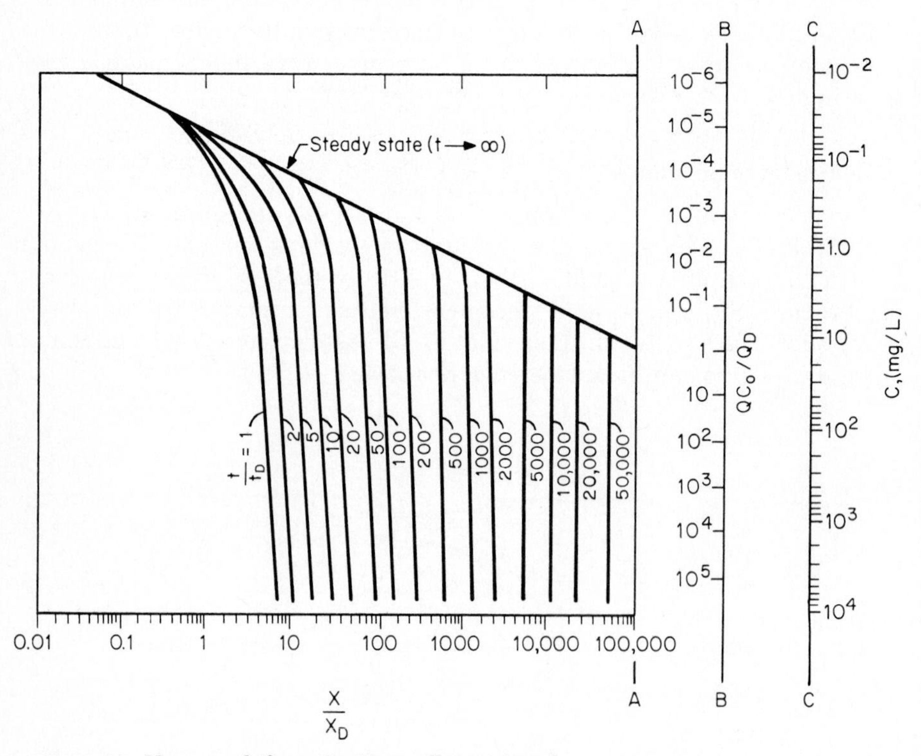

Figure 3.2 Nomograph for estimating pollutant travel.

- Pollutant concentration at source, C_0, mg/L
- Background concentration in groundwater C_b, mg/L
- Mass flow rate of contaminant, QC_0, kg/day

Use of the nomograph requires the calculation of three scale factors:

$$X_D = \frac{D_x}{v} = a_x \tag{3.7}$$

$$t_D = \frac{R_d D_x}{v^2} \tag{3.8}$$

$$Q_D = 16.02nz(D_x D_y)^{1/2} \tag{3.9}$$

The procedure is best illustrated with an example.

Example 3.1 Determine the nitrate concentration in the centerline of the plume, 600 m down gradient of an RI system, 2 years after system start-up. Data: aquifer thickness = 5 m, porosity = 0.35, seepage velocity = 0.45 m/day, dispersivity a_x = 32 m, a_y = 6 m, volumetric flow rate = 90 m³/day, nitrate concentration in percolate = 20 mg/L, nitrate concentration in background groundwater = 4 mg/L.

solution

1. The down-gradient volumetric flow rate combines the natural background flow plus the additional water introduced by the RI system. To be conservative, assume for this calculation that the total nitrate at the origin of the plume is equal to the specified 20 mg/L. The residual concentration determined with the nomograph is then added to the 4 mg/L background concentration to determine the total down-gradient concentration at the point of concern. Experience has shown that nitrate tends to be a conservative substance once the percolate passes the active root zone in the soil, so for this case assume that the retardation factor R_d is equal to 1.
2. Determine the dispersion coefficients:

$$D_x = a_x v = (32)(0.45) = 14.4 \text{ m}^2/\text{day}$$

$$D_y = a_y v = (6)(0.45) = 2.7 \text{ m}^2/\text{day}$$

3. Calculate the scale factors:

$$X_D = \frac{D_x}{v} = \frac{14.4}{0.45} = 32 \text{ m}$$

$$t_D = \frac{R_d D_x}{v^2} = \frac{(1)(14.4)}{(0.45)^2} = 71 \text{ days/m}$$

$$Q_D = 16.02nz(D_x D_y)^{1/2}$$
$$= (16.02)(0.35)(5)[(14.4)(2.7)]^{1/2} = 174.8 \text{ kg/day}$$

4. Determine the mass flow rate of the contaminant:

$$Q_{C_6} = \frac{(90 \text{ m}^3/\text{day})(20 \text{ mg/L})}{(1000 \text{ g/kg})} = 1.8 \text{ kg/day}$$

5. Determine the entry parameters for the nomograph:

$$\frac{x}{x_D} = \frac{600}{32} = 18.8$$

$$\frac{t}{t_D} = \frac{(2)(365)}{71} = 10.3 \text{ (use } t/t_D = 10 \text{ curve)}$$

$$\frac{Q_{C_0}}{Q_D} = \frac{1.8}{174.8} = 0.01$$

6. Enter the nomograph on the x/x_D axis with the value of 18.8, and draw a vertical line to intersect with the t/t_D 10 = curve. From that point project a line horizontally to the *A*-*A* axis. Locate the calculated value 0.01 on the *B*-*B* axis, connect this with the previously determined point on the *A*-*A* axis, extend this line to the *C*-*C* axis, and read the concentration of concern, which is about 0.4 mg/L.

7. The nitrate concentration at the 600 m point down gradient after 2 years is the sum of the nomograph value and the background concentration, or 4.4 mg/L.

Calculations must be repeated for each contaminant with use of the appropriate retardation factor. The nomograph can also be used to estimate the distance at which a given concentration will be present in a given time. The upper line on the figure is the steady-state curve for very long time periods and, as shown in Example 3.2, can be used to evaluate conditions when equilibrium is reached.

Example 3.2 Using the data from the previous example, determine the distance down gradient at which the groundwater in the plume will satisfy the Environmental Protection Agency (EPA) limits for nitrate in drinking water supplies (10 mg/L).

solution

1. Assuming a 4 mg/L background value, the plume concentration at the point of concern could be as much as 6 mg/L. Locate 6 mg/L on the *C*-*C* axis.

2. Connect the point on the *C*-*C* axis with the 0.01 value on the *B*-*B* axis (as determined in the previous example). Extend this line to the *A*-*A* axis. Project a horizontal line from this point to intersect the steady-state line. Project a vertical line downward to the x/x_D axis and read a value of $x/x_D = 60$.

3. Calculate the distance x using the previously determined value for x_D:

$$x = (x_D)(60) = (32)(60) = 1920 \text{ m}$$

Groundwater mounding

Groundwater mounding is illustrated schematically in Fig. 3.1. The percolate flow in the unsaturated zone is essentially vertical and controlled by K_v. If a groundwater table, impeding layer, or barrier exists at depth, a horizontal component is introduced and flow is controlled by a combination of K_v and K_h within the groundwater mound. At the margins of the mound and beyond, the flow is typically lateral and K_h controls.

The capability for lateral flow away from the source will determine the extent of mounding that will occur. The zone available for lateral flow includes the underground aquifer plus whatever additional elevation is considered acceptable for the particular project design. Excessive mounding will inhibit infiltration in an RI system. As a result the capillary fringe above the groundwater mound should never be closer than about 0.6 m (2 ft) to the infiltration surfaces in the RI basins. This will correspond to a water table depth of about 1 to 2 m (3 to 7 ft), depending on the soil texture.

In many cases the percolate or plume from an RI system will emerge as base flow in adjacent surface waters, and it may be necessary to estimate the position of the groundwater table between the source and the point of emergence. Such an analysis will reveal if seeps or springs are likely to develop in the intervening terrain. In addition, there are some regulatory agencies that require a specific residence time in the soil to protect adjacent surface waters, so it may be necessary to calculate the travel time from the source to the expected point of emergence. Equation 3.9 can be used to estimate the saturated thickness of the water table at any point down gradient of the source.[35] Typically, the calculation is repeated for a number of locations and the results are converted to an elevation and plotted on maps and profiles to identify potential problem areas.

$$h = \left(h_0^2 - \frac{2Q_i d}{K_h}\right)^{1/2} \tag{3.9}$$

where h = saturated thickness of the unconfined aquifer at the point of concern, m (ft)

h_0 = saturated thickness of the unconfined aquifer at the source, m (ft)

d = lateral distance from the source to the point of concern, m (ft)

K_h = effective horizontal permeability of the soil system, m/day (ft/day)

Q_i = lateral discharge from the unconfined aquifer system per unit width of the flow system, $m^3/(day \cdot m)$ $[ft^3/(day \cdot ft)]$

$$Q_i = \frac{K_h}{2d_i}(h_0^2 - h_i^2) \tag{3.10}$$

where d_i = distance to the seepage face or outlet point, m (ft)

h_i = saturated thickness of the unconfined aquifer at the outlet point, m (ft)

The travel time for lateral flow is a function of the hydraulic gradient, the distance traveled, the K_h, and the porosity of the soil as defined by Eq. 3.11.

$$t_D = \frac{nd_i^2}{K_h(h_0 - h_i)} \tag{3.11}$$

where t_D = travel time for lateral flow from source to the point of emergence in surface waters, m (ft)
K_h = effective horizontal permeability of the soil system, m/day (ft/day)
h_0 and h_1 = saturated thickness of the unconfined aquifer at the source and the outlet point, respectively, m (ft)
d_i = distance to the seepage face or outlet point, m (ft)
n = porosity

A simplified graphical method for determining groundwater mounding uses the procedure developed by Glover[14] and summarized by Bianchi and Muckel.[5] The method is valid for square or rectangular basins that lie above level, fairly thick, homogeneous aquifers of assumed infinite extent. However, the behavior of circular basins can be adequately approximated by assuming a square of equal area. When groundwater mounding becomes a critical project issue, further analysis using the Hantush method[1] is recommended. Further complications arise with sloped water tables, impeding subsurface layers that induce "perched" mounds, and the presence of a nearby outlet point (Refs. 7, 17, and 34 are suggested for these conditions). The simplified method involves graphical determination of several factors from Fig. 3.3, 3.4, 3.5, or 3.6, depending on whether the basin is square or rectangular.

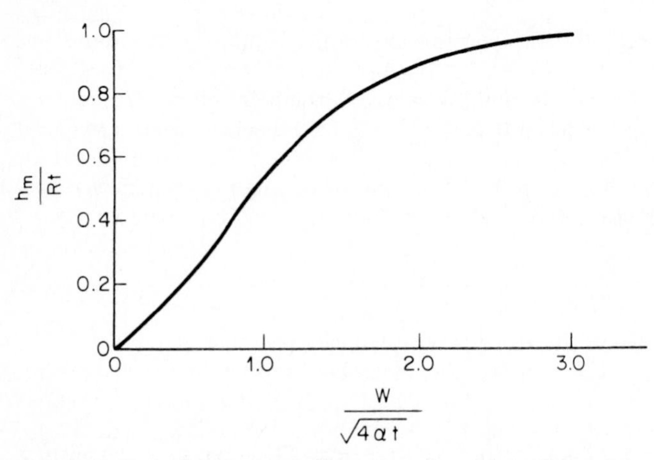

Figure 3.3 Groundwater mounding curve for center of a square recharge basin.

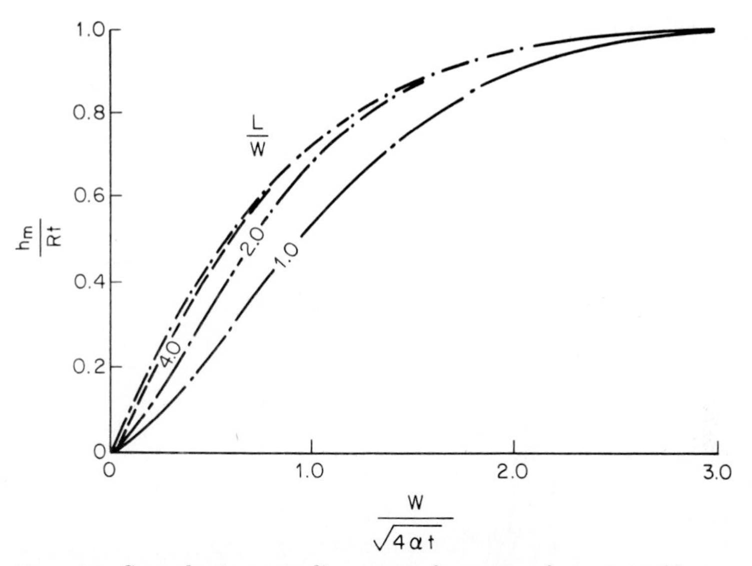

Figure 3.4 Groundwater mounding curves for center of a rectangular recharge area with different ratios of length L to width W.

It is necessary to calculate the values of $W/(4\alpha t)^{1/2}$ and of Rt, as defined in Eqs. 3.12 through 3.14:

$$\frac{W}{(4\alpha t)^{1/2}} = \text{dimensionless scale factor} \tag{3.12}$$

where W is the width of the recharge basin in meters (feet).

$$\alpha = \frac{K_h h_0}{Y_s} \tag{3.13}$$

where K_h = effective horizontal permeability of the aquifer, m/day (ft/day)
h_0 = original saturated thickness of the aquifer beneath the center of the recharge area, m (ft)
Y_s = specific yield of the soil (from Figs. 2.4 or 2.5), m^3/m^3 (ft^3/ft^3)

$$Rt = \text{scale factor, m (ft)} \tag{3.14}$$

where $R = I/Y_s$, m/day (ft/day)
I = infiltration rate (volume of water infiltrated per unit area of soil surface), $m^3/(m^2 \cdot day)$ [$ft^3/(ft^2 \cdot day)$]
t = period of infiltration, days

Either Fig. 3.3 or Fig. 3.4 is entered with the calculated value of $W/(4\alpha t)^{1/2}$ to determine the value for the ratio h_m/Rt, where h_m is the rise at the center of the mound. Using the previously calculated value for Rt, one can then solve for h_m.

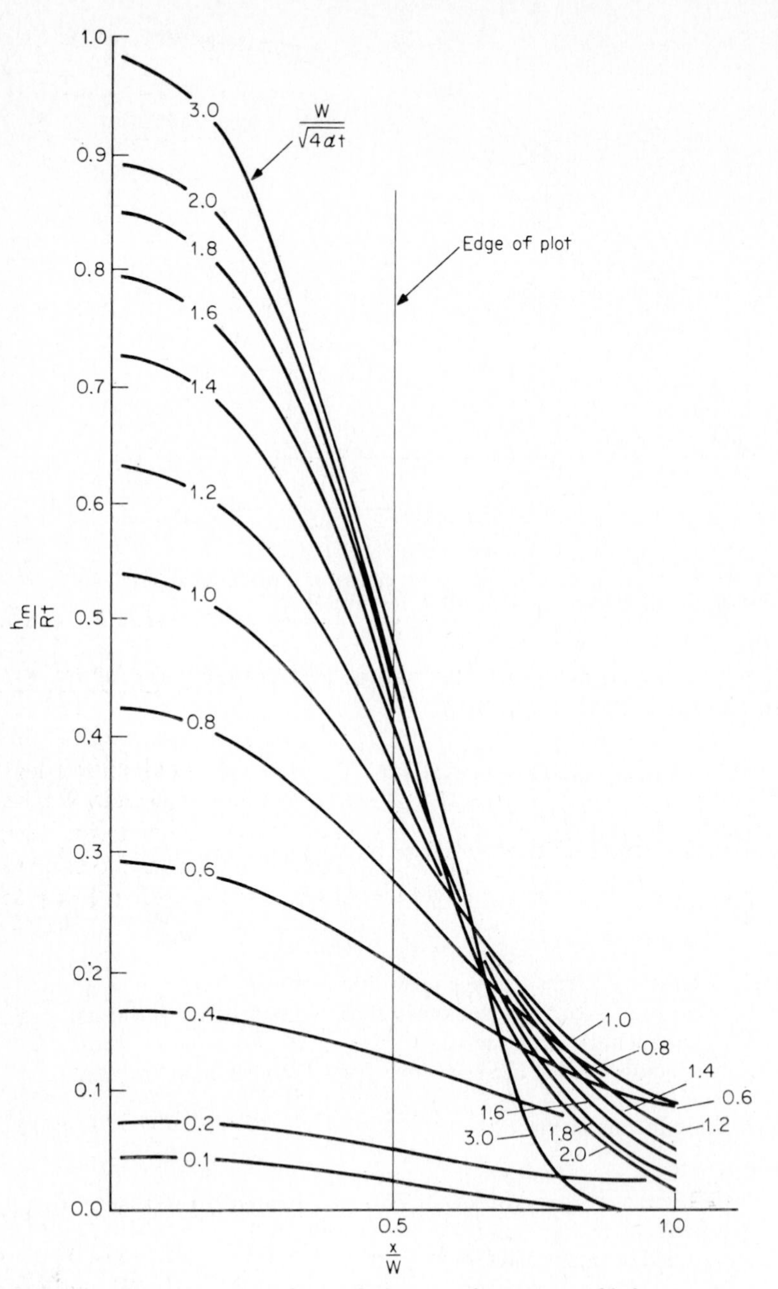

Figure 3.5 Rise and horizontal spread of a groundwater mound below a square recharge area.

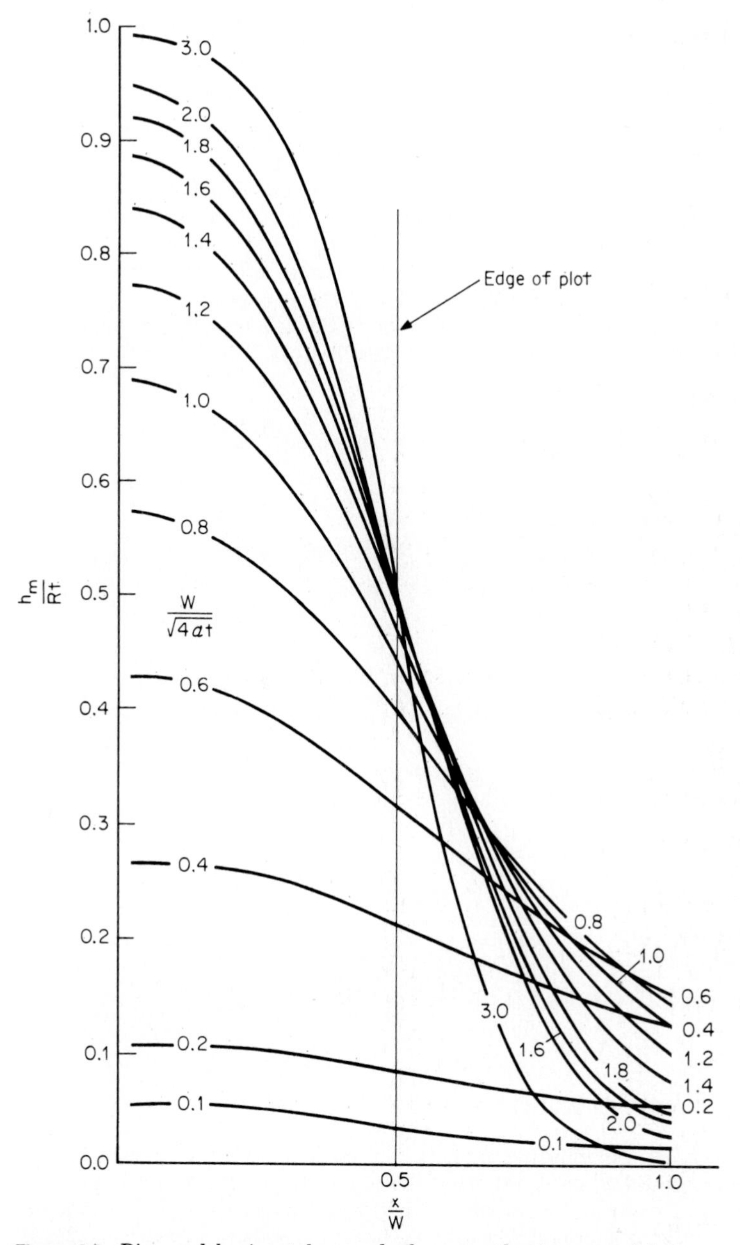

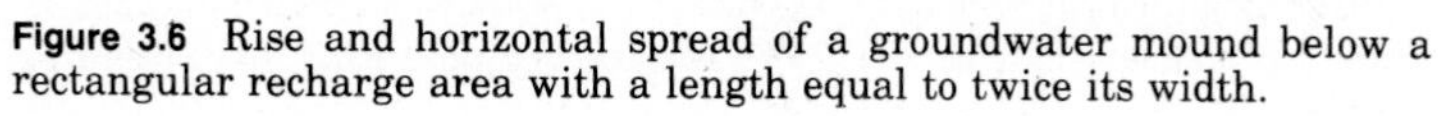

Figure 3.6 Rise and horizontal spread of a groundwater mound below a rectangular recharge area with a length equal to twice its width.

Figure 3.5 (for square areas) and Fig. 3.6 (for rectangular areas where $L = 2W$) are used to estimate the depth of the mound at various distances from the center of the recharge area. The procedures involved are best illustrated with a design example.

Example 3.3 Determine the height and the horizontal spread of a groundwater mound beneath a circular RI basin 30 m in diameter. The original aquifer thickness is 4 m, and the K_h as determined in the field is 1.25 m/day. The top of the original groundwater table is 6 m below the design infiltration surface of the constructed basin. The design infiltration rate will be 0.3 m/day, and the wastewater application period will be 3 days in every cycle (3 days of flooding, 10 days for percolation and drying; see Chap. 7 for details).

solution

1. Determine the size of an equivalent area square basin.

$$A = \frac{3.14D^2}{4} = 706.5 \text{ m}$$

Width W of equivalent square basin $= (706.5)^{1/2} = 26.6$ m

2. Use Fig. 2.5 to determine specific yield Y_s.

$$K_h = 1.25 \text{ m/day} = 5.21 \text{ cm/h}$$

$$Y_s = 0.14$$

3. Determine the scale factors:

$$\alpha = \frac{K_h h_0}{Y_s} = \frac{(1.25)(4)}{0.14} = 35.7 \text{ m}^2\text{/day}$$

$$W/(4\alpha t)^{1/2} = \frac{26.6}{[(4)(35.7)(3)]^{1/2}} = 1.28$$

$$R = \frac{0.3}{0.15} = 2 \text{ m/day}$$

$$Rt = (2)(3) = 6 \text{ m}$$

4. Use Fig. 3.3 to determine the factor h_m/Rt

$$\frac{h_m}{Rt} = 0.68$$

$$h_m = (0.68)(R)(t) = (0.68)(2)(3) = 4.08 \text{ m}$$

5. The original groundwater table is 6 m below the infiltration surface. The calculated rise of 4.08 m would bring the top of the mound within 2 m of the basin infiltration surface. As discussed previously, this is just adequate to maintain design infiltration rates. The designer might consider a shorter (say, 2-day) flooding period, as discussed in Chap. 7, to reduce the potential for mounding somewhat.

6. Use Fig. 3.5 to determine the lateral spread of the mound. Use the curve for $W/(4\alpha t)^{1/2}$ with the previously calculated value of 1.28, enter the graph with selected values of x/W (where x = lateral distance of concern), and read values of h_m/Rt.

Find depth to top of mound 10 m from centerline of basin:

$$\frac{x}{W} = \frac{10}{26.5} = 0.377$$

Enter x/W axis with this value, project up to $W/(4\alpha t)^{1/2} = 1.28$, then read 0.58 on h_m/Rt axis.

$$h_m = (0.58)(2)(3) = 3.48 \text{ m}$$

Depth to mound at 10 m point = 6 m − 3.48 m = 2.52 m

Similarly: at x = 13 m, depth to mound = 3.72 m
at x = 26 m, depth to mound = 5.6 m

This indicates that for this case the water level is almost back to the normal groundwater level at a lateral distance about equal to twice the basin width. Changing the application schedule to 2 days instead of 3 would reduce the peak water level to about 3 m below the infiltration surface of the basin.

The procedure demonstrated in Example 3.3 is valid for a single basin. However, as described in Chap. 7, RI systems typically include multiple basins that are loaded sequentially, and it is not appropriate to do the mounding calculation by assuming that the entire treatment area is uniformly loaded at the design hydraulic loading rate. In many situations this will result in the erroneous conclusion that mounding will interfere with the system operation.

It is necessary to first calculate the rise in the mound beneath a single basin during the flooding period. When hydraulic loading stops at time t, a uniform, hypothetical discharge, starting at t and continuing for the balance of the rest period, is assumed. The algebraic sum of these two mound heights then approximates the mound shape just prior to the start of the next flooding period. Since adjacent basins may be flooded during this same period, it is necessary to also determine the lateral extent of their mounds and then to add any increment from these sources in order to determine the total mound height beneath the basin of concern. The procedure is illustrated by Example 3.4.

Example 3.4 Determine the groundwater mound height beneath an RI basin at the end of the operational cycle. Assume that the basin is square, 26.5 m on a side, and is one in a set of four, arranged in a row (forming a rectangle 26.5 m wide, 106 m long). Assume the same site conditions used in Example 3.3. Also assume that flooding commences in one of the adjacent basins as soon as the rest period for the basin of concern starts. The operational cycle is 2 days flood, 12 days rest.

solution

1. The maximum rise beneath the basin of concern would be the same as that calculated in Example 3.3 with a 2-day flooding: h_m = 3.00 m.
2. The influence from the next 2 days of flooding in the adjacent basin would be about equal to the mound rise at the 26 m point calculated in Example 3.3,

or 0.4 m. All the other basins are beyond the zone of influence so the maximum potential rise beneath the basin of concern is:

$$3.00 + 0.4 = 3.4 \text{ m}$$

The mound will actually not rise that high, since during the 3 days that the adjacent basin is being flooded the first basin is draining. However, for the purposes of this calculation assume that the mound will rise the entire 3.4 m above the static groundwater table.

3. The R value for this "uniform" discharge will be the same as that calculated in Example 3.2, but t will now be 12 days

$$Rt = (2)(12) = 24 \text{ m/day}$$

4. Calculate a new $W/(4\alpha t)^{1/2}$ since the new time is 12 days.

$$W/(4\alpha t)^{1/2} = \frac{26.6}{[(4)(35.7)(12)]^{1/2}} = 0.62$$

5. Use Fig. 3.3 to determine $h_m/Rt = 0.30$

$$h_m = (24)(0.3) = 7.2 \text{ m}$$

This is the hypothetical drop in the mound that could occur during the 10-day rest period. However, the water level cannot actually drop below the static groundwater table so the maximum possible drop would be 3.4 m. This indicates that the mound would dissipate well before the start of the next flooding cycle. Assuming the drop to be at a uniform rate of 0.72 m/day, the 3.4 m mound would be gone in 4.7 days.

In cases in which the groundwater mounding analysis indicates potential interference with system operation, there are a number of corrective options. As described in Chap. 7, the flooding and drying cycles can be adjusted or the layout of the basin sets rearranged in a configuration with less interbasin interference. The final option is to underdrain the site to physically control the mound development.

Underdrainage may also be required to control shallow or seasonal natural groundwater levels when such levels might interfere with the operation of either land or aquatic treatment systems. Underdrains are also sometimes used to recover the treated water beneath land treatment systems for beneficial use or discharge elsewhere.

Underdrainage

In order to be effective, drainage or water recovery elements must be either at or within the natural groundwater table or just above some other flow barrier. When drains can be installed at depths of 5 m (16 ft) or less, underdrains are more effective and less costly than a series of wells. Modern techniques make it possible to rapidly install semi-flexible plastic drain pipe enclosed in a geotextile membrane by using a single machine that will cut and then close the trench.

In some cases underdrains are a project necessity to control a shallow groundwater table so that the site may be developed for wastewater treatment. Such drains, if effective for groundwater control will also collect the treated percolate from the land treatment operation. The collected water must be discharged, so the use of underdrains in this case converts the project to a surface water discharge system unless the water is otherwise used or disposed of. In a few situations drains have been installed to control a seasonally high water table. This type of system may require a surface water discharge permit during the period of high groundwater but will function as a nondischarging system for the balance of the year.

The drainage design consists of selecting the depth and spacing for placement of the drainpipes or tiles. In the typical case drains may be at a depth of 1 to 3 m (3 to 10 ft) and spaced 60 m (200 ft) or more apart. In sandy soils the spacing may approach 150 m (500 ft). The closer spacings provide better water control but the costs increase significantly.

The Hooghoudt[20] method is the most commonly used method for calculating the drain spacing. The procedure assumes that the soil is homogeneous, that the drains are spaced evenly apart, that Darcy's law is applicable, that the hydraulic gradient at any point is equal to the slope of the water table above that point, and that a barrier of some type underlies the drain. Figure 3.7 defines the necessary parameters for drain design, and Eq. 3.15 can be used for design.

$$S = \left[\frac{4K_h h_m}{L_w + P}(2d + h_m)\right]^{1/2} \tag{3.15}$$

where S = drain spacing, m (ft)
K_h = horizontal permeability of the soil, m/day (ft/day)
h_m = height of groundwater mound above the drains, m (ft)
L_w = annual wastewater loading rate expressed as a daily rate, m/day (ft/day)
P = average annual precipitation expressed as a daily rate, m/day (ft/day)
d = distance from drain to barrier, m (ft)

The position of the top of the mound between the drains is established by design or regulatory requirements for a particular project. For example, RI systems need a few meters of unsaturated soil above the mound in order to maintain the design infiltration rate, and SR systems also require an unsaturated zone to provide desirable conditions for the surface vegetation. (See Chap. 7 for further detail; procedures and criteria for more complex drainage situations can be found in Refs. 32 and 39.)

3.2 Biodegradable Organics

Biodegradable organic contaminants in either the dissolved or suspended form are characterized by the *biochemical oxygen demand* (BOD)

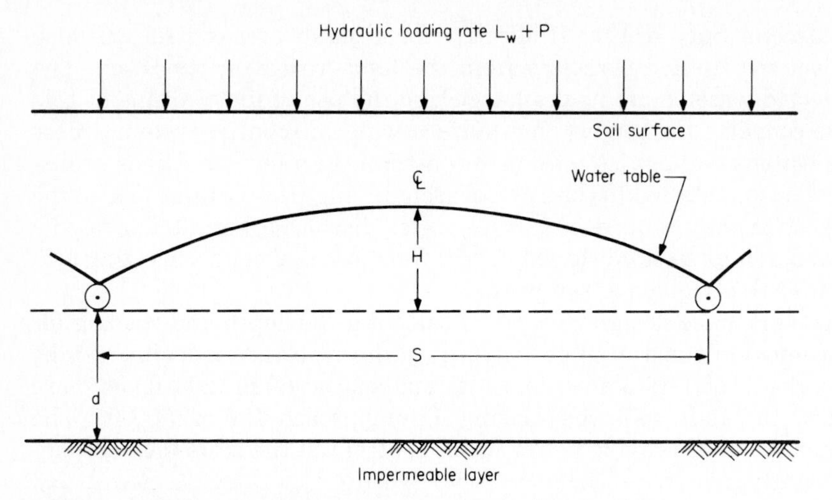

Figure 3.7 Definition sketch for calculation of drain spacing.

of the waste. Tables 1.1 through 1.3 present typical BOD removal expectations for the natural treatment systems described in this book.

Removal of BOD

As explained in Chaps. 4, 5, and 6, the BOD loading can be the limiting design factor for pond, aquaculture, and wetland systems. The basis for these limits is the maintenance of aerobic conditions within the upper water column in the unit and the resulting control of odors. The natural sources of dissolved oxygen in these systems are surface reaeration and photosynthetic oxygenation. Surface reaeration can be significant under windy conditions or if surface turbulence is created by mechanical means. Observation has shown that the dissolved oxygen in unaerated wastewater ponds varies almost directly with the level of photosynthetic activity, being low at night and early morning and rising to a peak in the early afternoon. The photosynthetic responses of algae are controlled by the presence of light, the temperature of the liquid, and the availability of nutrients and other growth factors.

Because algae are difficult to remove and can represent an unacceptable level of suspended solids in the effluent, some pond and aquaculture processes utilize mechanical aeration as the oxygen source. In partially mixed aerated ponds the increased depth of the pond and the partial mixing of the somewhat turbid contents limits the development of algae as compared with a facultative pond. Hyacinth ponds (Chap. 5) and many wetland systems (Chap. 6) restrict algae growth since the vegetation limits the penetration of light to the water column.

Both hyacinths and the emergent plant species used in wetlands

TABLE 3.2 Typical Organic Loading Rates for Natural Treatment Systems

Process	Organic loading, kg/(ha · day)
Oxidative pond	40–120
Facultative pond	22–67
Aerated, partial mix pond	50–200
Hyacinth ponds	20–50
Constructed wetlands*	< 120
Slow rate land treatment	50–500
Rapid infiltration land treatment	145–1000
Overland flow land treatment	40–110
Land application of municipal sludge	27–930†

* The free water surface type; see Chap. 6 for details.
† These values determined by dividing the annual rate by 365 days.

treatment have the unique capability to transmit oxygen from the leaf to the plant root. These plants do not themselves directly remove the BOD; they rather serve as hosts for a variety of attached growth organisms, and it is this microbial activity that is primarily responsible for the organic decomposition. The extensive root system of the hyacinth plant and the stems, stalks, roots, and rhizomes of the emergent varieties provide the necessary surfaces. This dependence then requires a relatively shallow reactor and a relatively low flow velocity to ensure optimum contact opportunities between the wastewater and the attached microbial growth.

The BOD of the wastewater or sludge is seldom the limiting design factor for the land treatment processes described in Chaps. 7 and 8. Some other factor, such as nitrogen, metals, toxics, or the hydraulic capacity of the soils, controls the design, so the system almost never approaches the upper limits for successful biodegradation of organics. Table 3.2 presents typical organic loadings for natural treatment systems.

Suspended solids removal

The suspended solids content of wastewater is not usually a limiting factor for design, but the improper management of solids within the system can result in process failure. One critical concern for both aquatic and terrestrial systems is the attainment of proper distribution of solids within the treatment reactor. The use of inlet diffusers in ponds, step feed (multiple inlets) in wetland channels, and higher-pressure sprinklers on industrial *overland flow* (OF) systems are all intended to achieve a more uniform distribution of solids and the avoidance of anaerobic conditions at the head of the process.

The removal of suspended solids in pond systems depends primarily on gravity sedimentation, and as mentioned previously, algae can be a concern in some situations. Sedimentation and entrapment in the microbial growths are both contributing factors in hyacinth ponds, wetlands, and the OF process. Filtration in the soil matrix is the principal mechanism for SR and RI systems. Removal expectations for the various processes are listed in Tables 1.1 through 1.3. The removal will typically exceed secondary treatment levels except for some of the pond systems that contain algal solids in their effluents.

3.3 Stable Organics

Stable organic materials, also called *refractory organics,* are more resistant to biological decomposition. Some are almost totally resistant and may persist in the environment for considerable periods of time; others are toxic or hazardous and require special management.

Removal methods

Volatilization, adsorption, and then biodegradation are the principal methods for removal of trace organics in natural treatment systems. Volatilization can occur at the water surface of ponds, wetlands, and RI basins, in the water droplets from sprinklers used in land treatment, from the liquid films in OF systems, and from the exposed surfaces of sludge. Adsorption occurs primarily on the organic matter in the treatment system that is in contact with the waste. In many cases microbial activity will then degrade the adsorbed materials.

Volatilization. The loss of volatile organics from a water surface can be described with first-order kinetics since it is assumed that the concentration in the atmosphere above the water surface is essentially zero. Equation 3.16 is the basic kinetic equation, and Eq. 3.17 can be used to determine the half-life for the contaminant of concern (see Chap. 8 for further discussion of the half-life concept and application to sludge organics).

$$\frac{C_t}{C_0} = \exp(-k_{vol}t/y) \tag{3.16}$$

where C_t = concentration at time t, mg/L (or μg/L)
C_0 = initial concentration at $t = 0$, mg/L (or μg/L)
k_{vol} = volatilization mass transfer coefficient, cm/h
$= ky$
k = overall rate coefficient, h^{-1}
y = depth of liquid, cm

$$t_{1/2} = \frac{0.693y}{k_{vol}} \tag{3.17}$$

where $t_{1/2}$ is time at which concentration $C_t = \frac{1}{2}C_0$, h.

The volatilization mass transfer coefficient is a function of the molecular weight of the contaminant and the air-water partition coefficient defined by the Henry's law constant, as shown by Eq. 3.18.

$$k_{vol} = \frac{B_1}{y} \frac{H}{(B_2 + H)M^{1/2}} \tag{3.18}$$

where k_{vol} = volatilization coefficient, h^{-1}
H = Henry's law constant, 10^5 (atm · m³)/mol
M = molecular weight of contaminant of concern, g/mol

The coefficients B_1 and B_2 are specific to the physical system of concern. Dilling[11] determined values for a variety of volatile chlorocarbons at a well-mixed water surface:

$$B_1 = 2.211 \qquad B_2 = 0.01042$$

Jenkins et al.[16] experimentally determined values for a number of volatile organics on an OF slope:

$$B_1 = 0.2563 \qquad B_2 = 5.86 \times 10^{-4}$$

The coefficients for the OF case are much lower since the flow of liquid down the slope is nonturbulent and could be considered almost laminar (Reynolds number = 100 to 400). The average depth of flowing liquid on this slope was about 1.2 cm.[16]

Using a variation of Eq. 3.18, Parker and Jenkins[22] determined volatilization losses from the droplets at a low-pressure large-droplet wastewater sprinkler. In this case the y term in the equation is equal to the average droplet radius; as a result their coefficients are only valid for the particular sprinkler system used. The approach is valid and can be evaluated for other sprinklers and operating pressures. Equation 3.19 was developed by Parker and Jenkins for the organic compounds listed in Table 3.3.

$$\ln \frac{C_t}{C_0} = 4.535(k_{vol'} + 11.02 \times 10^{-4}) \tag{3.19}$$

TABLE 3.3 Volatile Organic Removal by Wastewater Sprinkling[22]

Substance	Calculated $k_{vol'}$ for Eq. 3.19, cm/min	Substance	Calculated $k_{vol'}$ for Eq. 3.19, cm/min
Chloroform	0.188	Hexane	0.239
Benzene	0.236	Nitrobenzene	0.0136
Toluene	0.220	*m*-Nitrotoluene	0.0322
Chlorobenzene	0.190	PCB 1242	0.0734
Bromoform	0.0987	Naphthalene	0.114
m-Dichlorobenzene	0.175	Phenanthrene	0.0218
Pentane	0.260		

Volatile organics can also be removed by aeration in pond systems. Clark et al.[8] developed Eq. 3.20 to determine the amount of air required to strip a given quantity of volatile organics from water via aeration.

$$(A/W) = 76.4\left(1 - \frac{C_t}{C_0}\right)^{12.44} S^{0.37} V^{-0.45} M^{-0.18} 0.33^s \qquad (3.20)$$

where (A/W) = air/water ratio
S = solubility of organic compound, mg/L
V = vapor pressure, mmHg
M = molecular weight, g/mol
s = saturation of the compound of concern
= 0 for unsaturated organics, 1 for saturated compounds

The values in Table 3.4 can be used in Eq. 3.20 to calculate the air/water ratio required for some typical volatile organics.

Adsorption. Sorption of trace organics by the organic matter present in the treatment system is thought to be the primary physicochemical mechanism for removal. The concentration of the trace organic that is sorbed relative to that in solution is defined by a partition coefficient K_p, which is related to the solubility of the chemical. This value can be estimated if the octanol-water partition coefficient K_{ow} and the percentage of organic carbon in the system are defined, as shown by Eq. 3.21.

$$\log K_{oc} = 1.00 \log K_{ow} - 0.21 \qquad (3.21)$$

where K_{oc} = sorption coefficient expressed on an organic carbon basis
= $K_{sorb}/(OC)$
K_{sorb} = sorption mass transfer coefficient, cm/h (in/h)
OC = percentage of organic carbon present in the system
K_{ow} = octanol/water partition coefficient

Reference 15 presents other correlations and a detailed discussion of sorption in soil systems.

TABLE 3.4 Properties of Selected Volatile Organics For Eq. 3.20[19]

Chemical	M	S	s
Trichloroethylene	132	1000	0
1,1,1-Trichloroethane	133	5000	1
Tetrachloroethylene	166	145	0
Carbon tetrachloride	154	800	1
cis-1,2-Dichloroethylene	97	3500	0
1,2-Dichloroethane	99	8700	1
1,1-Dichloroethylene	97	40	0

Jenkins et al.[16] determined that sorption of trace organics on an OF slope could be described by first-order kinetics with the rate constant defined by Eq. 3.22.

$$k_{sorb} = \frac{B_3}{y} \times \frac{K_{ow}}{(B_4 + K_{ow})M^{1/2}} \tag{3.22}$$

where k_{sorb} = sorption coefficient, h^{-1}
B_3 = coefficient specific to the treatment system
= 0.7309 for the OF system studied
y = depth of water on slope (~1.2 cm)
K_{ow} = octanol-water partition coefficient
B_4 = coefficient specific to the treatment system
= 170.8 for the OF system studied
M = molecular weight of the organic chemical, g/mol

In many cases the removal of trace organics is due to a combination of sorption and volatilization. The overall process rate constant k_{sv} is then the sum of the coefficients defined with Eqs. 3.18 and 3.22, and the combined removal described by Eq. 3.23.

$$\frac{C_t}{C_0} = \exp(-k_{sv}t) \tag{3.23}$$

where k_{sv} = overall rate constant for combined volatilization and sorption
= $k_{vol} + k_{sorb}$
C_t = concentration at time t, mg/L (or μg/L)
C_0 = initial concentration at $t = 0$, mgL (or μg/L)

TABLE 3.5 Physical Characteristics for Selected Organic Chemicals

Substance	K_{ow}*	H†	Vapor pressure‡	M§
Chloroform	93.3	314	194	119
Benzene	135	435	95.2	78
Toluene	490	515	28.4	92
Chlorobenzene	692	267	12.0	113
Bromoform	189	63	5.68	253
m-Dichlorobenzene	2.4×10^3	360	2.33	147
Pentane	1.7×10^3	125,000	520	72
Hexane	7.1×10^3	170,000	154	86
Nitrobenzene	70.8	1.9	0.23	123
m-Nitrotoluene	282	5.3	0.23	137
Diethyl phthalate	162	0.056	7×10^{-4}	222
PCB 1242	3.8×10^5	30	4×10^{-4}	26
Naphthalene	2.3×10^3	36	8.28×10^{-2}	128
Phenanthrene	2.2×10^4	3.9	2.03×10^{-4}	178
2,4-Dinitrophenol	34.7	0.001	—	184

* Octanol-water partition coefficient.
† Henry's law constant, 10^5 atm · m³/mol at 20°C and 1 atm.
‡ At 25°C.
§ Molecular weight, g/mol.

Table 3.5 presents the physical characteristics of a number of volatile organics for use in the equations presented above for volatilization and sorption.

Example 3.5 Determine the removal of toluene on an OF land treatment system. Assume 30-m long terrace, hydraulic loading of 0.4 $m^3/(hr \cdot m)$ (see Chap. 7 for discussion), mean residence time on slope is 90 min, wastewater application with a low-pressure large-droplet sprinkler, physical characteristics for toluene are $K_{ow} = 490$, $H = 515$, $M = 92$ (Table 3.5), depth of flowing water on the terrace is 1.5 cm, and concentration of toluene in applied wastewater is 70 μg/L.

solution

1. Use Eq. 3.19 to estimate volatilization losses during sprinkling.

$$\ln \frac{C_t}{C_0} = 4.535(k_{vol'} + 11.02 \times 10^{-4})$$

$$C_t = 70 \exp\left[-(4.535)(0.220 + 0.001102)\right]$$

$$= 25.69 \ \mu g/L$$

2. Use Eq. 3.18 to determine the volatilization coefficient during flow on the OF terrace.

$$k_{vol} = \frac{B_1}{y} \frac{H}{(B_2 + H)M^{1/2}}$$

$$= \frac{0.2563}{1.5} \frac{515}{(5.86 \times 10^{-4} + 515)92^{1/2}}$$

$$= (0.17087)(0.1042)$$

$$= 0.0178$$

3. Use Eq. 3.22 to determine the sorption coefficient during flow on the OF terrace.

$$k_{sorb} = \frac{B_3}{y} \frac{K_{ow}}{(B_4 + K_{ow})M^{1/2}}$$

$$k_{sorb} = \frac{0.7309}{1.5} \frac{490}{(170.8 + 490)92^{1/2}}$$

$$= (0.4873)(0.0774)$$

$$= 0.0377$$

4. The overall rate constant is the sum of k_{vol} and k_{sorb}.

$$k_t = 0.0178 + 0.0377 = 0.0555$$

5. Use Eq. 3.23 to determine the toluene concentration in the OF runoff.

$$\frac{C_t}{C_0} = \exp(-k_t t)$$

$$C_t = 25.68 \exp\left[-(0.0555)(90)\right]$$

$$= 0.17 \ \mu g/L$$

This represents about 99.8 percent removal.

Removal performance

The land treatment systems are the only natural treatment concepts that have been studied extensively to determine the removal of the

TABLE 3.6 Removal of Organic Chemicals in Land Treatment Systems

Substance	SR*		OF†, %	RI‡, %
	Sandy soil, %	Silty soil, %		
Chloroform	98.57	99.23	96.50	> 99.99
Toluene	> 99.99	> 99.99	99.00	99.99
Benzene	> 99.99	> 99.99	98.09	> 99.99
Chlorobenzene	99.97	99.98	98.99	> 99.99
Bromoform	99.93	99.96	97.43	> 99.99
Dibromochloromethane	99.72	99.72	98.78	> 99.99
m-Nitrotoluene	> 99.99	> 99.99	94.03	—§
PCB 1242	> 99.99	> 99.99	96.46	> 99.99
Naphthalene	99.98	99.98	98.49	96.15
Phenanthrene	> 99.99	> 99.99	99.19	—
Pentachlorophenol	> 99.99	> 99.99	98.06	—
2,4-Dinitrophenol	—	—	93.44	—
Nitrobenzene	> 99.99	> 99.99	88.73	—
m-Dichlorobenzene	> 99.99	> 99.99	—	82.27
Pentane	> 99.99	> 99.99	—	—
Hexane	99.96	99.96	—	—
Diethyl phthalate	—	—	—	90.75

* Ref. 22.
† Ref. 16.
‡ Ref. 19.
§ Not reported.

priority pollutant organic chemicals. This is probably due to the greater concern for groundwater contamination with these systems. Results from these studies have been generally positive. As indicated previously, the more soluble compounds such as chloroform tend to move more rapidly through the soil system than the less soluble materials such as some PCBs. In all cases the amount escaping the treatment system with percolate or effluent is very small. Table 3.6 presents removal performance for the three major land treatment concepts. The removals observed in the SR system were after 1.5 m of vertical travel in the soils indicated, and a low-pressure large-droplet sprinkler was used for the application. The removals on the OF system were measured after flow on a terrace about 30 m long, with application via gated pipe at the top of the slope at a hydraulic loading of 0.12 $m^3/(m \cdot h)$. The RI data were obtained from wells about 200 m down gradient of the application basins.

The removals reported in Table 3.6 for SR systems represent concentrations in the applied wastewater ranging from 2 to 111 μg/L and percolate concentrations ranging from 0 to 0.4 μg/L. The applied concentrations on the OF system ranged from 25 to 315 μg/L and effluent concentrations from 0.3 to 16 μg/L. Concentrations of the reported substances applied to the RI system ranged from 3 to 89 μg/L and percolate concentrations from 0.1 to 0.9.

The results in Table 3.6 indicate that the SR system was more consistent and had higher removals than the other two systems. This is probably due to the use of the sprinkler and the enhanced opportunity for sorption on the organic matter in these finer-textured soils. Chloroform was the only compound to consistently appear in the percolate, and that was at very low concentrations. Although slightly less effective than the SR system, the other two methods still produced very high removals. If sprinklers had been used on the OF system, it is likely that the removals would have been even higher. Based on these data it would appear that all three land treatment techniques will be more effective for trace organic removal than activated sludge and other conventional mechanical treatment systems.

Quantitative relationships have not yet been developed for trace organic removal from natural aquatic systems. The removal due to volatilization in pond and free-water-surface wetland systems can at least be estimated with Eqs. 3.18 and 3.23. The liquid depth in these systems is much greater than on an OF slope, but the detention time is measured in terms of many days instead of minutes, so the removal can still be very significant. Organic removal in the submerged-bed wetlands may be comparable with the RI values in Table 3.6, depending on the media used in the wetlands.

Travel time in soils

The rate of movement of organic compounds in soils is a function of the velocity of the carrier water, the organic content of the soil, the octanol-water partition coefficient for the organic compound, and other physical properties of the soil system. Equation 3.24 can be used to estimate the movement velocity of an organic compound during saturated flow in the soil system.

$$V_c = \frac{KG}{n - 0.63p(\mathrm{OC})K_{\mathrm{ow}}} \tag{3.24}$$

where V_c = velocity of organic compound, m/day (ft/day)
K = saturated permeability of soil, m/day (ft/day), in vertical or horizontal direction
K_v = saturated vertical permeability, m/day (ft/day)
K_h = saturated horizontal permeability, m/day (ft/day)
G = hydraulic gradient of flow system, m/m (ft/ft)
= 1 for vertical flow
= $\Delta H/\Delta L$ for horizontal flow, m/m (ft/ft) (see Eq. 3.4 for definition)
n = porosity of the soil (as a decimal fraction; see Fig. 2.4)
p = bulk density of soil, g/cm^3 (lb/in^3)
OC = organic carbon content of soil, decimal fraction
K_{ow} = octanol-water partition coefficient

3.4 Pathogens

Pathogenic organisms may be present in both wastewaters and sludges, and their control is one of the fundamental reasons for waste management. Bacterial limits on discharges to surface waters are specified by many regulatory authorities. Other potential risks are impact on groundwater from both aquatic and land treatment systems, contamination of crops or infection of grazing animals on land treatment sites, and offsite loss of aerosolized organisms from pond aerators or land treatment sprinklers. Investigations[25] have shown that the natural aquatic, wetland, and land treatment concepts provide very effective control of pathogens.

Aquatic systems

The removal of pathogens in pond type systems is due to natural die-off, predation, sedimentation, and adsorption. Helminths, *Ascaris,* and other parasitic cysts and eggs will settle to the bottom in the quiescent zone of ponds. Facultative ponds with three cells and about 20 days detention time and aerated ponds with a separate settling cell prior to discharge will provide more than adequate helminth and protozoa removal. As a result there is little risk of parasitic infection from pond effluents or from use of such effluents in agriculture. There may be some risk when sludges are removed for disposal. Either these sludges can be treated or temporary restrictions on public access and agricultural use may be placed on the disposal site.

Bacteria and virus removal. The removal of both bacteria and viruses in multiple cell pond systems of either the aerated or unaerated type is very effective, as shown in Tables 3.7 and 3.8.

TABLE 3.7 Fecal Coliform Removal in Pond Systems[37]

Location	No. of cells	Detention time, days	Fecal coli, no./100 mL	
			Influent	Effluent
Facultative Ponds				
Peterborough, N.H.	3	57	4.3×10^6	3.6×10^5
Eudora, Kan.	3	47	2.4×10^6	2.0×10^2
Kilmichael, Miss.	3	79	12.8×10^6	2.3×10^4
Corinne, Utah	7	180	1.0×10^6	7.0×10^0
Partial Mix Aerated Ponds				
Windber, Pa.	3	30	10^6	3.0×10^2
Edgerton, Wis.	3	30	10^6	3.0×10^1
Pawnee, Ill.	3	60	10^6	3.3×10^1
Gulfport, Miss.	2	26	10^6	1.0×10^5

TABLE 3.8 Enteric Virus Removal in Facultative Ponds[2]

Location	Enteric virus PFU/L*	
	Influent	Effluent
Shelby, Miss., 3 cells, 72 days		
Summer	791	0.8
Winter	52	0.7
Spring	53	0.2
El Paso, Tex., 3 cells, 35 days		
Summer	348	0.6
Winter	87	1.0
Spring	74	1.1
Beresford, S.D., 2 cells, 62 days		
Summer	94	0.5
Winter	44	2.2
Spring	50	0.4

* PFU/L = Plaque-forming units per liter.

The effluent in all three cases described in Table 3.8 was undisinfected. The viruses measured were the naturally occurring enteric types and not seeded virus or bacteriophage. Table 3.8 presents seasonal averages; Ref. 27 contains full details. The viral concentrations in the effluent were consistently low at all times, although as shown in Table 3.8, the removal efficiency did drop slightly in the winter in all three locations.

Numerous studies have shown that the removal of fecal coliforms in ponds is dependent on detention time and temperature; it can be estimated by Eq. 3.25, in which the detention time used is the actual detention time in the system as measured by dye studies. The actual detention time in a pond can be as little as 45 percent of the theoretical design detention time owing to short circuiting of flow. If dye studies are not practical or possible, it would be conservative to assume that the actual detention time is 50 percent of the design residence time for use in Eq. 3.25.

$$\frac{C_f}{C_i} = \frac{1}{(1 + tk_T)^n} \tag{3.25}$$

where C_i = influent fecal coli concentration, number per 100 mL
C_f = effluent fecal coli concentration, number per 100 mL
t = actual detention time in the cell, days
n = number of cells in series
k_T = temperature-dependent rate constant, days^{-1}
= $(2.6)(1.19)^{(T_w - 20)}$
T_w = mean water temperature in pond, °C

See Chap. 4 for a method for determining temperature in a pond. For the general case it is safe to assume that the water temperature will be about equal to the mean monthly air temperature, down to a minimum of 2°C.

Equation 3.25 in the form presented assumes that all cells in the system have the same size (see Chap. 4 for the general form of the equation when the cells are of different sizes). The equation can be rearranged and solved to determine the optimum number of cells needed for a particular level of pathogen removal. In general, a three- or four-cell (series) system with an actual detention time of about 20 days will remove fecal coliforms to desired levels. Model studies with polio- and coxsackie viruses indicated that the removal of viruses proceeds similarly to the first-order reaction described by Eq. 3.25. Hyacinth ponds and similar aquatic units should perform in accordance with Eq. 3.25 also.

Wetland systems

Pathogen removal in many wetland systems is due to essentially the same factors described above for pond systems. Equation 3.25 can also be used to estimate the removal of bacteria or virus in wetland systems where the water flow path is above the surface. The detention time will be less in most constructed wetlands as compared with ponds, but the opportunities for adsorption and filtration will be greater. The wetland systems described in Chap. 6, which depend on subsurface lateral flow, will remove pathogens in essentially the same ways as land treatment systems. Table 3.9 summarizes information on pathogen removal from selected wetlands.

Land treatment systems

Since land treatment systems in the United States are typically preceded by some form of preliminary treatment and/or a storage pond, there should be little concern with parasites. The infection of grazing animals is due to the direct ingestion of essentially raw wastewater.[25] The removal of bacteria and viruses in land treatment systems is due to a combination of filtration, dessication, adsorption, radiation, and predation.

Ground surface aspects. The major concerns relate to the potential for the contamination of surface vegetation or off-site runoff. The persistence of bacteria or viruses on plant surfaces could then infect humans or animals if the plants were consumed raw. It is generally recommended in the United States that in order to eliminate these risks

TABLE 3.9 Pathogen Removal in Constructed Wetland Systems[29–31,36]

Location	System performance	
	Influent	Effluent*
Santee, Calif., bullrush wetland†		
Winter season (Oct.–Mar.)		
Total coli, no./100 mL	5×10^7	1×10^5
Bacteriophage, PFU/mL	1900	15
Summer season (Apr.–Sept.)		
Total coli, no./100 mL	6.5×10^7	3×10^5
Bacteriophage, PFU/mL	2300	26
Iselin, Pa., cattails & grasses‡		
Winter season (Nov.–Apr.)		
Fecal coli, no./100 mL	1.7×10^6	6200
Summer season (May–Oct.)		
Fecal coli, no./100 mL	1.0×10^6	723
Arcata, Calif., bullrush wetland§		
Winter season		
Fecal coli, no./100 mL	4300	900
Summer season		
Fecal coli, no./100 mL	1800	80
Listowel, Ont., cattails§		
Winter season		
Fecal coli, no./100 mL	556,000	1400
Summer season		
Fecal coli, no./100 mL	198,000	400

* Undisinfected.
† Gravel bed, subsurface flow.
‡ Sand bed, subsurface flow.
§ Free water surface.

agricultural land treatment sites not grow vegetables to be eaten raw. The major risk will be to grazing animals on a pasture irrigated with wastewater. Typical criteria specify a period ranging from 1 to 3 weeks after sprinkling undisinfected effluent before grazing animals are allowed access. Systems of this type are divided into relatively small paddocks and the animals moved in rotation around the site.

Control of runoff is a design requirement of SR and RI land treatment systems, as described in Chap. 7, so there should be no pathogenic hazard from these sources. Runoff of the treated effluent is the design intention of OF systems, which typically can achieve about 90 percent removal of the applied fecal coliforms. The need for final disinfection of treated OF runoff will be a site-specific decision by the regulatory agency. The OF slopes also collect precipitation of any intensity that may occur. The runoff from these rainfall events can be more intense than the design treatment rate, but the additional dilution provided results in equal or better water quality than the normal runoff.

Groundwater contamination. Since percolate from SR and RI land treatment can reach groundwater aquifers, the risk of pathogenic contamination must be considered. The removal of bacteria and viruses in the finer-textured agricultural soils used in SR systems is quite effective. A 5-year study in Hanover, N.H.[25] demonstrated almost complete removal of fecal coliforms within the top 5 ft of the soil profile. Similar studies in Canada[4] indicated that the fecal coliforms were retained in the top 8 cm (3 in) of the soil. About 90 percent of the bacteria died within the first 48 h, and the remainder were eliminated over the next 2 weeks. Virus removal, which initially depends on adsorption, is also very effective on these soils.

The coarse-textured soils and high hydraulic loading rates used in RI systems increase the risk of bacteria and virus transmission to groundwater aquifers. A considerable research effort, both in the laboratory and at operational systems, has focused on viral movement in RI systems.[25] The results of this work indicate minimal risk for the general case; movement can occur with very high viral concentrations if the wastewater is applied at very high loading rates on very coarse-textured soils. It is unlikely that all three factors will be present in the majority of cases. Chlorine disinfection prior to wastewater application in an RI system is not recommended since the chlorinated organic compounds formed represent a greater threat to the groundwater than does the potential transmission of a few bacteria or viruses.

Sludge systems

As shown by the values in Table 3.10, the pathogen levels in raw and digested sludge can be quite high.

The pathogen content of sludge is especially critical when the sludge is to be used in agricultural operations or when public exposure is a concern. Two levels of stabilization are recognized by the Environmental Protection Agency. The first group includes aerobic and anaerobic digestion, air drying (for 3 months), lime stabilization, and composting [at 40°C (104°F), for 5 days] as well as other methods capable of the same results. The basic criteria are a 90 percent reduction

TABLE 3.10 Typical pathogen levels in wastewater sludges

Pathogen	Untreated, no./100 mL	Anaerobically digested, no./100 mL
Virus	2500–70,000	100–1000
Fecal coliforms	10×10^6	30,000–600,000
Salmonella	8000	3–62
Ascaris lumbricoides	200–1000	0–1000

in the pathogens present or a 99 percent reduction in fecal coliforms. Sludges with these characteristics would require limits on public access to the site for at least 12 months and limits on grazing animals for at least 1 month, with no contact between the sludge and edible portion of the crop permitted.

A higher level of stabilization is required if crops intended for direct human consumption are to be grown within 18 months of the sludge application or if it is desired to relax the constraints given in the previous paragraph. This next level of stabilization is termed *processes to further reduce pathogens* (PFRP). Recognized techniques include heat drying, heat treatment, thermophilic aerobic digestion, and composting at higher temperatures for longer periods than in the previous case. Pasteurization and irradiation are also acceptable following the basic stabilization procedures. Composting is described in detail in Chap. 8. Sludge stabilization with earthworms (vermistabilization) is also described in Chap. 8, and there is some evidence that a reduction in pathogenic bacteria occurs during the process. The freeze dewatering process will not kill pathogens but can reduce the concentration in the remaining sludge owing to enhanced drainage upon thawing. The reed bed drying concept can achieve significant pathogen reduction due to dessication and the long detention time in the system.

The pathogens are further reduced after sludge is land-applied by the same mechanisms discussed previously for land application of wastewater. There is little risk of transmission of sludge pathogens to groundwater or in runoff to surface waters if the criteria in Chap. 8 are used for system design.

Aerosols

Aerosol particles are 20 μm in diameter or less, which is still large enough to transport bacteria or virus. Aerosols will be produced whenever liquid droplets are sprayed into the air, as well as at the boundary layer above agitated water surfaces, or when sludges are moved about or aerated. Aerosol particles can travel significant distances, and the contained pathogens remain viable until inactivated by desiccation or ultraviolet light. The downwind travel distance for aerosol particles is dependent on wind speed, turbulence, temperature, humidity, and the presence of any barrier that might entrap the particle. With the impact sprinklers commonly used in land application of wastewater, the volume of aerosols produced amounts to about 0.3 percent of the water leaving the nozzle.[29] If a barrier is not present, the greatest travel distance will occur with steady nonturbulent winds under cool, humid conditions, which are generally most likely at night.

TABLE 3.11 Organism Concentration in Wastewater and Downwind Aerosol[30]

Organism	Wastewater concentration, (no./100 mL) × 10^6	Aerosol concentration at edge of sprinkler impact circle, no./m^3 of air sampled
Standard plate count	69.9	2578
Total coliform	7.5	5.6
Fecal coliform	0.8	1.1
Coliphage	0.22	0.4
Fecal streptococci	0.007	11.3
Pseudomonas	1.1	71.7
Klebsiella	0.39	< 1.0
Clostridium perfringes	0.005	1.4

The concentration of organisms entering a sprinkler nozzle should be no different from the concentration in the bulk liquid or sludge. Immediately after aerosolization, the temperature, sunlight, and humidity changes have an immediate and significant effect on the organism concentration. This *aerosol shock* is demonstrated in Table 3.11.

As the aerosol particle travels downwind, the microorganisms continue to die off at a slower, first-order rate as a result of desiccation, ultraviolet radiation, and possibly trace compounds in the air or in the aerosol. This die-off can be very significant for bacteria, but the rates for virus are very slow, so it is prudent to assume no further downwind inactivation of virus by these factors. Equations 3.26 and 3.27 represent a predictive model, which can be used to estimate the downwind concentration of aerosol organisms.

$$C_d = C_n D_d \exp(xa) + B \tag{3.26}$$

where C_d = concentration at distance d, no./m^3 (no./ft^3)
C_n = concentration released at source, no./s
D_d = atmospheric dispersion factor, s/m^3 (s/ft^3)
x = decay or die-off rate, s^{-1}
= −0.023 for bacteria (derived for fecal coliforms)
= 0.00 for virus (assumed)
a = downwind distance d/wind velocity, m/(m · s) [ft/(ft · s)]
B = background concentration in upwind air, no./m^3 (no./ft^3)

The initial concentration C_n leaving the nozzle area is a function of the original concentration in the bulk wastewater W, the wastewater flow rate F, the aerosolization efficiency E, and a survival factor I, all as described by Eq. 3.27.

$$C_n = WFEI \tag{3.27}$$

where C_n = organisms released at source, no./m^3 (no./ft^3)
W = concentration in bulk wastewater, no./100 mL
F = flow rate, L/s (0.631 gal/min)
E = aerosolization efficiency
= 0.003 for wastewater
= 0.0004 for sludge spray guns
= 0.000007 for sludge applied with tank truck sprinklers
I = survival factor
= 0.34 for total coliforms
= 0.27 for fecal coliforms
= 0.71 for coliphage
= 3.6 for fecal streptococci
= 80.0 for enteroviruses

The atmospheric dispersion factor D_d in Eq. 3.26 depends on a number of related meteorological conditions. Typical values for a range of expected conditions are given below; Ref. 36 should be consulted for a more exact determination.

Field condition	D_d, s/m^3
Wind < 6 km/h, strong sunlight	176×10^{-6}
Wind < 6 km/h, cloudy daylight	388×10^{-6}
Wind 6–16 km/h, strong sunlight	141×10^{-6}
Wind 6–16 km/h, cloudy daylight	318×10^{-6}
Wind > 16 km/h, strong sunlight	282×10^{-6}
Wind > 16 km/h, cloudy daylight	600×10^{-6}
Wind > 11 km/h, night	600×10^{-6}

The example below illustrates the use of this predictive model.

Example 3.6 Find the fecal coliform concentration in aerosols 8 m downwind of the sprinkler impact zone if the sprinkler has a 23-m impact circle and is discharging at 30 L/s; fecal coliforms in the bulk wastewater are 1×10^5; the sprinkler is operating on a cloudy day with a wind speed of about 8 km/h; and background concentration of fecal coliforms in the upwind air is zero.

solution

1. The distance of concern is 31 m downwind of the nozzle source, and the wind velocity is 2.22 m/s, so calculate the a factor:

$$a = \frac{\text{downwind distance}}{\text{wind velocity}} = \frac{31}{2.22} = 13.96\ \text{s}^{-1}$$

2. Calculate the concentration leaving the nozzle area with Eq. 3.27

$$C_n = WFEI$$

$$C_n = (1 \times 10^5)(30\ \text{L/s})(0.003)(0.27)$$

= 2430 fecal coliforms released per second at nozzle

3. Calculate the concentration at the downwind point of concern with Eq. 3.26.

$D_d = 318 \times 10^{-6}$

$C_d = C_n D_d \exp(xa) + B$

$= 2340(318 \times 10^{-6})\exp[-0.023(13.96)] + 0.0$

$= 0.54$ fecal coliforms per cubic meter of air, 8 m downwind of the wetted zone of the sprinkler. This concentration is an insignificant level of risk.

The very low concentration predicted in the example is typical of the very low concentrations actually measured at a number of operational land treatment sites. Table 3.12 is a summary of data collected from an intensively studied system in which undisinfected effluent was applied to the land.

It seems clear that the very low aerosolization efficiencies E, as defined in Eq. 3.26 for sludge spray guns and truck-mounted sprinklers, means that there is very little risk of aerosol transport of pathogens from these sources; this has been confirmed in field investigations.[31]

Composting is a very effective process for the inactivation of most microorganisms, including viruses, owing to the high temperatures generated during the treatment (see Chap. 8 for details). However, the heat produced in the process also stimulates the growth of thermophilic fungi and actinomycetes, and concerns have been expressed regarding their aerosol transport. The aerosols in this case are dust particles released when the compost materials are aerated, mixed, screened, or otherwise moved about the site.

A study was conducted at four composting operations involving 400 on-site and off-site workers.[9] The most significant finding was a higher concentration of the fungus *Aspergillus fumigatus* in the throat and nasal cultures of the actively involved on-site workers, but this finding

TABLE 3.12 Aerosol Bacteria and Virus at Pleasanton, California, Land Treatment System Using Undisinfected Effluent

Location	Fecal coliform	Fecal streptococci	Coliphage	*Pseudomonas*	Entroviruses
Wastewater (no./100 mL)	1×10^5	8.8×10^3	2.6×10^5	2.6×10^5	2.8
Upwind (no./m^3)	0.02	0.23	0.01	0.03	ND*
Downwind (no./m^3)					
10–30 m	0.99	1.45	0.34	81	0.01
31–80 m	0.46	0.60	0.39	46	ND
81–200 m	0.23	0.42	0.21	25	ND

* ND = None detected.

was not correlated with an increased incidence of infection or disease. The fungus was rarely detected in on-site workers only occasionally involved or in the off-site control group.

The presence of this fungus is due to the composting process itself and not to the fact that wastewater sludges were involved. The study results suggest that workers directly and frequently involved with the composting operations have a greater risk of exposure but that there is a negligible impact on those only occasionally exposed or on the downwind off-site population. It should be possible to protect all concerned with respirators for the exposed workers and a boundary screen of vegetation around the site.

3.5 Metals

Metals at trace level concentrations are found in all wastewaters and sludges. Industrial and commercial activities are the major sources, but wastewater from private residences can also have significant metal concentrations. The metals of greatest concern are copper, nickel, lead, zinc, and cadmium, and the reason for the concern is the risk of their entry into the food chain or water supply.

A large percentage of the metals present in wastewater will accumulate in the sludges produced during the wastewater treatment process. As a result, metals are often the controlling design parameter for land application of sludge, as described in detail in Chap. 8. Metals are not usually the critical design parameter for wastewater treatment or reuse, with the possible exception of certain industrial wastes. Table 3.13 compares the metal concentrations in untreated municipal wastewaters with the requirements for irrigation and drinking water supplies.

TABLE 3.13 Metal Concentrations in Wastewater and Requirements for Irrigation and Drinking Water Supplies[34]

Metal	Untreated* wastewater, mg/L	Drinking water, mg/L	Irrigation, mg/L	
			Continuous†	Short-term‡
Cadmium	< 0.005	0.01	0.01	0.05
Lead	0.008	0.05	5.0	10.0
Zinc	0.04	0.05	2.0	10.0
Copper	0.18	1.0	0.2	5.0
Nickel	0.04	—	0.2	2.0

* Median values for typical municipal wastewater.

† For waters used for an indefinite time on any kind of soil.

‡ For waters used for up to 20 years on fine-textured soils when sensitive crops are to be grown.

Aquatic systems

Trace metals are not usually a concern for the design or performance of pond systems treating typical municipal wastewaters. The major pathways for removal are adsorption on organic matter and precipitation. Since the opportunity for both is somewhat limited, the removal of metals in most pond systems will be less effective than, for example, in activated sludge systems, where more than 50 percent of the metals present in the untreated wastewater can be transferred to the sludge in a relatively short time period. Sludges from pond systems can, however, contain relatively high concentrations of metals owing to the long retention times and the infrequent sludge removals. The metal concentrations found in lagoon sludges at several locations are summarized in Table 3.14.

The concentrations shown in Table 3.14 are within the range normally found in unstabilized primary sludges and therefore would not inhibit further digestion or land application as described in Chap. 8. (Tables 8.4 and 8.5 list other characteristics of pond sludges.) The data in Table 3.14 are from lagoons in cold climates. It is likely that sludge metal concentrations may be higher than these values in warm climate lagoons that receive a significant industrial wastewater input. In these cases the benthic sludge will undergo further digestion, which reduces the organic content and sludge mass, but the metals remain unaffected, so their concentration should increase with time.

TABLE 3.14 Metal Concentrations in Sludges from Treatment Lagoons[28]

Metal	Facultative lagoons*	Partial-mix aerated lagoons†
Copper		
Wet sludge, mg/L	3.8	10.1
Dry solids, mg/kg	53.8	809.2
Iron		
Wet sludge, mg/L	0.1	1.2
Dry solids, mg/kg	9.0	9.2
Lead		
Wet sludge, mg/L	8.9	21.1
Dry solids, mg/kg	144	394
Mercury		
Wet sludge, mg/L	0.1	0.2
Dry solids, mg/kg	2.4	4.7
Zinc		
Wet sludge, mg/L	54.6	85.2
Dry solids, mg/kg	840	2729

* Average of values from two facultative lagoons in Utah.
† Average of values from two partial-mix aerated lagoons in Alaska.

TABLE 3.15 Metal Removal in Hyacinth Ponds[18]

Metal	Influent concentration, μg/L	Percent removal*
Boron	140	37
Copper	27.6	20
Iron	457.8	34
Manganese	18.2	37
Lead	12.8	68
Cadmium	0.4	46
Chromium	0.8	22
Arsenic	0.9	18

* Average of three parallel channels, detention time about 5 days.

If metals removal is a process requirement and the local climate is close to subtropical, the use of water hyacinths in shallow ponds, as described in Chap. 5, may be considered. Tests with full-scale systems in both Louisiana and Florida[18] have documented excellent removal, with uptake by the plant itself a major factor. The plant tissue concentrations could range from hundreds to thousands of times that of the water or sediment concentrations, which indicates that bioaccumulation of trace elements by the plant occurs. Metal removals in a pilot hyacinth system in central Florida are presented in Table 3.15.

Hyacinths have also been shown to be particularly effective in extracting metals from photoprocessing wastewater at a system in Louisiana.[18]

Wetland systems

Excellent metal removals have been demonstrated in the type of constructed wetlands described in Chap. 6. Tests at pilot wetlands in southern California, with about 5.5 days hydraulic residence time, indicated 99, 97, and 99 percent removal for copper, zinc, and cadmium, respectively.[13] However, plant uptake by the vegetation accounted for less than 1 percent of the metals involved. The major mechanisms responsible for metal removal were precipitation and adsorption interactions with the organic benthic layer.

Land treatment systems

Removal of metals in land treatment systems can involve both uptake by any vegetation present and adsorption, ion exchange, precipitation, and complexation in or on the soil. As explained in Chap. 8, zinc, copper, and nickel will be toxic to the vegetation long before they reach a concentration in the plant tissue that would represent a risk to human or animal food chains. Cadmium, however can accumulate in many plants without toxic effects and may represent some health risk. As a result, cadmium is the major limiting factor for application of sludge on agricultural land.

The near-surface soil layer in land treatment systems is very effective for removal, and most retained metals are found in this zone. Investigations at a rapid infiltration system that had operated for 33 years on Cape Cod, Massachusetts[25] indicated that essentially all the metals applied could be accounted for in the top 50 cm (20 in) of the sandy soil, and over 95 percent were contained within the top 15 cm (6 in).

Although the metal concentrations in typical wastewaters are low, concerns have been expressed regarding long-term accumulation in the soil, which might then affect the future agricultural potential of the site. Work by Hinesly and others[25] seems to indicate that most of the metals retained over a long period in the soil are in forms that are not readily available to most vegetation. Plants will respond to the metals applied during the current growing season but are not significantly affected by the previous accumulations in the soil. The data in Table 3.16 demonstrate the same relationship. In Melbourne, Australia after 76 years of application of raw sewage, the cadmium concentration in the grass on the site is just slightly higher than in the grass on the control site, which received no wastewater. The other locations are newer systems in California, and the cadmium content is of the same order of magnitude as that measured at Melbourne, suggesting that the vegetation in all these locations is responding to the metals applied during the current growing season and not to the prior soil accumulation. The significantly higher lead content of the three California sites as compared with Melbourne is believed due to motor vehicle exhausts operating on adjacent highways.

Metals do not pose a threat to groundwater aquifers, even at the very high hydraulic loadings used in RI systems. Experience at Hol-

TABLE 3.16 Metal Content of Grasses at Land Treatment Sites (concentrations, mg/L)

	Locations: date started, date sampled				
	Melbourne				
Metal	Control	1896, 1972	Fresno, 1907, 1973	Manteca, 1961, 1973	Livermore, 1964, 1973
Cadmium	0.77	0.89	0.9	1.6	0.3
Copper	6.5	12.0	16.0	13.0	10.0
Nickel	2.7	4.9	5.0	45.0	2.0
Lead	2.5	2.5	13.0	15.0	10.0
Zinc	50.0	63.0	93.0	161.0	103.0

lister, California demonstrates that the concentration of cadmium in the shallow groundwater beneath the site is not significantly different from normal offsite groundwater quality. After 33 years of operation at this site the accumulation of metals in the soil is still below or near the low end of the range normally expected for agricultural soils. Had the site been operated in the SR mode, it would have taken over 150 years to apply the same volume of wastewater and contained metals.

3.6 Nutrients

There is a dual concern with respect to nutrients, since their control is necessary to avoid adverse health or environmental effects but the same nutrients are essential for the performance of the natural biological treatment systems discussed in this book. The nutrients of major importance for both purposes are nitrogen, phosphorus, and potassium. Nitrogen is the controlling parameter for the design of many land treatment and sludge application systems; those aspects are discussed in detail in Chaps. 7 and 8. This section covers the potential for nutrient removal in the other treatment methods and the nutrient requirements of the various system components.

Nitrogen

Nitrogen is limited in drinking water to protect the health of infants and may be limited in surface waters to protect fish or to avoid eutrophication. As described in Chap. 7, land treatment systems are typically designed to meet the 10 mg/L nitrate drinking water standard for any percolate or groundwater leaving the project boundary. In some cases nitrogen removal may also be necessary prior to discharge to surface waters. More often, there is a need to oxidize or otherwise remove nitrogen present in the form of ammonia, since this is toxic to many fish and can also represent a significant oxygen demand on the stream.

Nitrogen is present in wastewaters in a variety of forms because of the various oxidation states represented, and it can readily change from one state to another depending on the physical and biochemical conditions present. The total nitrogen concentration in typical municipal wastewaters ranges from about 15 to over 50 mg/L. About 60 percent of this is in the ammonia form and the remainder in the organic form.

Ammonia can be present as molecular ammonia, NH_3 or as ammonium ions, NH_4^+. The equilibrium between these two forms in water is strongly dependent on pH and temperature. At pH 7 essentially only ammonium ions are present and at pH 12 only dissolved ammonia gas.

This relationship is the basis for air-stripping operations in advanced wastewater treatment plants and for a significant portion of the nitrogen removal that occurs in wastewater treatment ponds.

Aquatic systems. Nitrogen can be removed in pond systems by plant or algal uptake, nitrification and denitrification, adsorption, sludge deposition, and loss of ammonia gas to the atmosphere (volatilization). In facultative wastewater treatment ponds the dominant mechanism is believed to be volatilization, and under favorable conditions up to 80 percent of the total nitrogen present can be lost. The rate of removal is dependent on pH, temperature, and detention time. The amount of gaseous ammonia present at near neutral pH levels is relatively low, but when some of this gas is lost to the atmosphere, additional ammonium ions shift to the ammonia form to maintain the equilibrium. Although the unit rate of conversion and loss may be very low, the long detention time in these ponds compensates, which results in a very effective removal over the long term. Equations, which can be used for design, describing this nitrogen removal in ponds are presented in Chap. 4. Since nitrogen is often the controlling design parameter for land treatment, a reduction in pond effluent nitrogen can often permit a very significant reduction in the land area needed for wastewater application, with a comparable savings in project costs.

Aquaculture systems. Nitrogen removal in hyacinth ponds can be very effective and is primarily due to nitrification/denitrification and plant uptake. The plant uptake will not represent permanent removal unless the plants are routinely harvested. A complete harvest is not typically possible, since another function of the hyacinth plant is to shade the water surface so that the restricted light penetration will limit algal growth. Since the harvest might only remove 20 to 30 percent of the plants in the basin at any one time, the full nitrogen removal potential of the plants is never realized.

Nitrification and denitrification are possible in shallow hyacinth ponds even if mechanical aeration is used, owing to the presence of aerobic and anaerobic microsites within the dense root zone of the floating plant and the presence of the carbon sources needed for denitrification. Nitrogen removals observed in hyacinth ponds range from less than 10 to over 50 kg/(ha · day) [9 to 45 lb/(acre · day)] depending on the season and the frequency of harvest. Some of these ponds were carefully managed pilot-scale or research facilities. See Chap. 5 for further discussion.

Wetland systems. Volatilization of ammonia, denitrification, and plant uptake (if the vegetation is harvested) are the potential methods of nitrogen removal in wetland systems. Studies in Canada[40] demonstrated that a regular harvest of cattails still only accounted for about 10 percent of the nitrogen removed by the system. These findings have been confirmed elsewhere, which indicates that the major pathway for nitrogen removal is nitrification/denitrification. A pilot system in California cuts and mulches the wetland vegetation in place to provide additional carbon and achieves about 98 percent nitrogen removal.[12]

Land treatment systems. Nitrogen is usually the limiting design parameter for SR land treatment of wastewater; criteria and procedures are presented in Chap. 7. Nitrogen can also limit the annual application rate for many sludge systems, as described in Chap. 8. The removal pathways for both types of systems are similar and include plant uptake, ammonia volatilization, and nitrification/denitrification.

Ammonium ions can be adsorbed onto soil particles, and this provides a temporary control; soil microorganisms then nitrify this ammonium, restoring the original adsorptive capacity. Nitrate, on the other hand, will not be chemically retained by the soil system. Nitrate removal by plant uptake or denitrification can only occur during the hydraulic residence time of the carrier water in the soil profile. The overall capability for nitrogen removal will be improved if the applied nitrogen is ammonia or in some other less well oxidized form.

Nitrification/denitrification is the major factor for nitrogen removal in RI systems, and crop uptake is a major method for both SR and OF systems. Volatilization and denitrification also occur with the latter two treatment systems and may account for 10 to over 50 percent of the applied nitrogen, depending on waste characteristics and application methods as described in Chap. 7. Design procedures based on nitrogen uptake of agricultural and forest vegetation can also be found in Chap. 7.

Phosphorus

Phosphorus has no known health significance but is the wastewater constituent most often associated with eutrophication of surface waters. Phosphorus in wastewater can occur as polyphosphates, as orthophosphates, which can originate from a number of sources, and as organic phosphorus, which is more commonly found in industrial discharges. The potential removal pathways in natural treatment systems include vegetation uptake, other biological processes, adsorption, and precipitation.

The vegetative uptake can be significant in the SR and OF land treatment processes when harvest and removal are routinely practiced. In these cases the harvested vegetation might account for 20 to 30 percent of applied phosphorus. The vegetation typically used in wetland systems is not considered a significant factor for phosphorus removal, even if harvesting is practiced. If the plants are not harvested, their decomposition releases phosphorus back to the water in the system. Phosphorus removal by water hyacinths and other aquatic plants is limited to the plant needs and will not exceed 50 to 70 percent of the phosphorus present in the wastewater, even with careful management and regular harvests.

Adsorption and precipitation reactions are the major pathways for phosphorus removal when wastewater has the opportunity for contact with a significant volume of soil. This is always the case with SR and RI systems and with some wetland systems, where infiltration and lateral flow through the subsoil are possible. The possibilities for contact between the wastewater and the soil are more limited with the OF concept, since relatively impermeable soils are used.

The soil reactions involve the clay, oxides of iron and aluminum, and calcium compounds that are present and the soil pH. Finer-textured soils tend to have the greatest potential for phosphorus sorption owing to the higher clay content but also to the increased hydraulic residence time. Coarse-textured, acidic, or organic soils have the lowest capacity for phosphorus. Peat soils are both acidic and organic, but some have a significant sorption potential due to the presence of iron and aluminum oxides.

A laboratory-scale adsorption test can estimate the amount of phosphorus that a soil can remove during short application periods. Actual phosphorus retention in the field will be at least 2 to 5 times the value obtained during a typical 5-day adsorption test. The sorption potential of a given soil layer will eventually be exhausted, but until that time occurs, the removal of phosphorus will be almost complete. It has been estimated that a 30-cm (11.8-in) depth of soil in a typical SR system might become saturated with phosphorus every 10 years. The phosphorus concentrations in the percolate from SR systems usually approach background levels for the native groundwater within 2 m (6.6 ft) of travel in the soil. The coarser-textured soils utilized for rapid infiltration might require an order-of-magnitude greater travel distance.

Phosphorus is not usually a critical factor for groundwater quality. However, when the groundwater emerges in a nearby surface stream or pond, there may be eutrophication concerns. Equation 3.28 can be used to estimate the phosphorus concentration at any point on the infiltration/percolation groundwater flow path. The equation was originally developed from RI system responses and so provides a very conservative basis for all soil systems.

$$P_x = P_0 \exp(-k_p t) \tag{3.28}$$

where P_x = total phosphorus at a distance x on the flow path, mg/L
P_0 = total phosphorus in applied wastewater, mg/L
k_p = 0.048 at pH 7, $days^{-1}$ (pH 7 gives the lowest value)
t = detention time, days
$= xW/K_xG$
x = distance along flow path, m (ft)
W = saturated soil water content, assume 0.4
K_x = hydraulic conductivity of soil in direction x, m/day (ft/day); K_v = vertical, K_h = horizontal
G = hydraulic gradient for flow system, $G = 1$ for vertical flow
$= \Delta H/\Delta L$ for lateral flow

The equation is solved in two steps, first for the vertical flow component from the soil surface to the subsurface flow barrier (if one exists) and then for the lateral flow to the adjacent surface water. The calculations are based on assumed saturated conditions, so the lowest possible detention time will result. The actual vertical flow in most cases will be under unsaturated conditions, so the actual detention time in this zone will be much longer than that calculated by this procedure. If the equation predicts acceptable removal, there is some assurance that the site will perform reliably, and detailed tests should not be necessary for preliminary work. Detailed tests should be conducted for final design of large-scale projects.

Potassium and other micronutrients

Potassium as a wastewater constituent usually has no health or environmental effects. It is, however, an essential nutrient for vegetative growth, and it is not typically present in wastewaters in the optimum combination with nitrogen and phosphorus. If a land or aquatic treatment system depends on vegetation for nitrogen removal, it may be necessary to add supplemental potassium to maintain plant uptake of nitrogen at the optimum level. Equation 3.29 can be used to estimate the supplemental potassium that may be required for aquatic systems and for land systems in which the soils have a low level of natural potassium.

$$K_s = 0.9U - K_{ww} \tag{3.29}$$

where K_s = annual supplemental potassium needed, kg/ha (lb/acre)
U = estimated annual nitrogen uptake of vegetation, kg/ha (lb/acre)
K_{ww} = amount of potassium in the applied wastewater, kg/ha (lb/acre)

Most plants also require magnesium, calcium, and sulfur, and depending on soil characteristics there may be deficiencies in some locations. Iron, manganese, zinc, boron, copper, molybdenum, and sodium are other micronutrients important for vegetative growth. In the general case there are sufficient amounts of these elements in wastewater, and in some cases the excess can lead to phytotoxicity problems. Some high-rate hyacinth systems may require supplemental iron to maintain vigorous plant growth.

Boron. Boron is at the same time essential for plant growth and toxic at low concentrations for sensitive plants. Experience has shown that soil systems have very limited capacity for boron adsorption, so it is conservative to assume a zero removal potential for land treatment systems. Industrial wastewaters can have a higher boron content than typical municipal effluents; the boron content may influence the type of crop selected but will not control the feasibility of land treatment. Tolerant crops such as alfalfa, cotton, sugar beets, and sweet clover might accept up to 2 to 4 mg/L boron in the wastewater. Semitolerant crops such as corn, barley, milo, oats, and wheat might accept 1 to 2 mg/L, and sensitive crops such as fruits and nuts should receive less than 1 mg/L.

Sulfur. Wastewaters contain sulfur in either the sulfite or sulfate form. Municipal wastewaters do not usually contain enough sulfur to be a design problem, but industrial wastewaters from petroleum refining and kraft paper mills can be a concern. Sulfate is limited to 250 mg/L in drinking waters and to 200 to 600 mg/L for irrigation, depending on the type of vegetation. Sulfur is weakly adsorbed on soils, so the major pathway for removal is by plant uptake. The grasses typically used in land treatment can remove 2 to 3 kg of sulfur per 1000 kg (4 to 7 lb per 2200 lb) of material harvested. The presence of sulfites or sulfates in wastewater can lead to serious odor problems if anaerobic conditions develop, as has occurred with some hyacinth systems. Supplemental aeration is then needed to maintain aerobic conditions in the basin.

Sodium. Sodium is not limited by primary drinking water standards nor are there significant environmental water quality concerns with respect to the sodium content of typical municipal wastewaters. A sudden change to a high sodium content will adversely affect the biota in aquatic systems, but most can acclimate to gradual changes. Sodium and also calcium influence soil alkalinity and salinity, which in turn can affect the vegetation in land treatment systems. The growth of the plant and its ability to absorb moisture from the soil are influenced by salinity.

The structure of clay soils can be damaged when there is an excess of sodium with respect to calcium and magnesium in the wastewater.

The resulting swelling of some clay particles changes the hydraulic capacity of the soil profile. The *sodium adsorption ratio* (SAR), as shown by Eq. 3.30, defines the relationship among these three elements.

$$\text{SAR} = \frac{\text{Na}}{\left(\dfrac{\text{Ca} + \text{Mg}}{2}\right)^{1/2}} \tag{3.30}$$

where SAR = sodium adsorption ratio
Na = sodium concentration, meq/L
= (mg/L in wastewater)/(22.99)
Ca = calcium concentration, meq/L
= (mg/L in wastewater)(2)/(40.08)
Mg = magnesium concentration, meq/L
= (mg/L in wastewater)(2)/(24.32)

The SAR for typical municipal effluents seldom exceeds a value of 5 to 8, so there should be no problem with most soils in any climate. Soils with up to 15 percent clay can tolerate an SAR of 10 or less, while soils with little clay or with nonswelling clays can accept SARs up to about 20. Industrial wastewaters can have a high SAR, and periodic soil treatment with gypsum or some other inexpensive source of calcium may be necessary to reduce the clay swelling.

Soil salinity is managed by adding an excess of water above that required for crop growth to leach the salts from the soil profile. A rule of thumb for total water needed to prevent salt build-up in arid climates is to apply the crop needs plus about 10 percent. Reference 35 provides further detail.

REFERENCES

1. Bauman, P.: "Technical Development in Ground Water Recharge," V. T. Chow (ed.), *Advances in Hydroscience,* Vol. 2, Academic Press, N.Y., 1965, pp. 209–279.
2. Bausum, H. T.: *Enteric Virus Removal in Wastewater Treatment Lagoon Systems,* PB83-234914, National Technical Information Service, Springfield, VA, 1983.
3. Bedient, P. B., N. K. Springer, E. Baca, T. C. Bouvette, S. R. Hutchins, and M. B. Tomson: "Ground-Water Transport From Wastewater Infiltration," *J. Environ. Eng.,* ASCE, 109(2):485–501, 1983.
4. Bell, R. G., and J. B. Bole: "Elimination of Fecal Coliform Bacteria from Soil Irrigated with Municipal Sewage Lagoon Effluent," *J. Environ. Quality,* 7:193–196, 1978.
5. Bianchi, W. C., and C. Muckel: *Ground Water Recharge Hydrology,* ARS 41-161, U.S. Dept. of Agriculture, Agricultural Research Station, Beltsville, Md., December 1970.
6. Bouwer, H.: *Groundwater Hydrology,* McGraw-Hill, New York, 1978.
7. Brock, R. P.: "Dupuit-Forchheimer and Potential Theories for Recharge from Basins," *Water Resources Res.,* 12:909–911, 1976.
8. Clark, C. S., H. S. Bjornson, J. Schwartz-Fulton, J. W. Holland, and P. S. Gartside: "Biological Health Risks Associated with the Composting of Wastewater Treatment Plant Sludge," *J. Water Pollution Control Fed.,* 56(12):1269–1276, 1984.

9. Clark, R. M., R. C. Eilers, and J. A. Goodrich: "VOCs in Drinking Water: Cost of Removal," *J. Environ. Eng.*, ASCE, 110(6):1146–1162, 1984.
10. Danel, P.: "The Measurement of Ground-Water Flow," *Proceedings of Ankara Symposium on Arid Zone Hydrology,* UNESCO, Paris, 1953, pp. 99–107.
11. Dilling, W. L.: "Interphase Transfer Processes. II. Evaporation of Chloromethanes, Ethanes, Ethylenes, Propanes, and Propylenes from Dilute Aqueous Solutions; Comparisons With Theoretical Predictions," *Environ. Sci. Technol.* 11:405–409, 1977.
12. Gersberg, R. M., B. V. Elkins, and C. R. Goldman: "Nitrogen Removal in Artificial Wetlands," *Water Res.,* 17(9):1009–1014, 1983.
13. Gersberg, R. M., S. R. Lyon, B. V. Elkins, and C. R. Goldman: "The Removal of Heavy Metals by Artificial Wetlands," *Proceedings American Water Works Assoc. Water Reuse Symp.,* American Water Works Assoc., Denver, Colo., 1985, pp. 639–645.
14. Glover, R. E.: *Mathematical Derivations as Pertaining to Groundwater Recharge,* U.S. Dept. of Agriculture, Agricultural Research Service, 81 pp., Beltsville, Md., 1961.
15. Hutchins, S. R., M. B. Tomsom, P. B. Bedient, and C. H. Ward: "Fate of Trace Organics During Land Application of Municipal Wastewater," *Crit. Rev. Environ. Control,* 15(4):355–416, CRC Press, Boca Raton, Fla., 1985.
16. Jenkins, T. F., D. C. Leggett, L. V. Parker, and J. L. Oliphant: "Toxic Organics Removal Kinetics in Overland Flow Land Treatment," *Water Res.,* 19(6):707–718, 1985.
17. Kahn, M. Y., and D. Kirkham: "Shapes of Steady State Perched Groundwater Mounds," *Water Resources Res.,* 12:429–439, 1976.
18. Kamber, D. M.: *Benefits and Implementation Potential of Wastewater Aquaculture,* EPA Contract Report 68-01-6232, Environmental Protection Agency, Office of Water Regulations and Standards, Washington, D.C., 1982.
19. Love, O. T., R. Miltner, R. G. Eilers, and C. A. Fronk-Leist: *Treatment of Volatile Organic Chemicals in Drinking Water,* EPA 600/8-83-019, U.S. Municipal Engineering Research Laboratory, Cincinnati, Ohio, 1983.
20. Luthin, J. N.: *Drainage Engineering,* Kreiger Publishing Co., Huntington, N.Y., 1973.
21. Overcash, M. R., and D. Pal: *Design of Land Treatment Systems for Industrial Wastes—Theory and Practice,* Ann Arbor Science, Ann Arbor, Mich., 1979.
22. Parker, L. V., and T. F. Jenkins: "Removal of Trace-Level Organics by Slow-Rate Land Treatment," *Water Res.,* 20(11):1417–1426, 1986.
23. Pettygrove, G. S., and T. Asano (eds): *Irrigation with Reclaimed Municipal Wastewater—A Guidance Manual,* prepared for California State Water Resources Control Board, reprinted by Lewis Publishers, Inc. Chelsea, Mich., 1985.
24. Pound, C. E., and R. W. Crites: Long Term Effects of Land Application of Domestic Wastewater—Hollister California, EPA 600/2-78-084, Environmental Protection Agency, Office of Research and Development, Washington, D.C., 1979.
25. Reed, S. C.: *Health Aspects of Land Treatment,* GPO 1979-657-093/7086, Environmental Protection Agency, Center for Environmental Research Information, Cincinnati, Ohio, 1979.
26. Reed, S., R. Bastian, S. Black, and R. Khettry: "Wetlands for Wastewater Treatment in Cold Climates," *Proceedings 3d American Water Works Assoc. Water Reuse Symp.,* American Water Works Assoc., Denver, Colo., 1985, pp. 962–972.
27. Roberts, P. V., P. L. McCarty, M. Reinhard, and J. Schriner: "Organic Contaminant Behavior During Groundwater Recharge," *J. Water Pollution Control Fed.,* 52(1):161–172, 1980.
28. Schneiter, R. W., and E. J. Middlebrooks: *Cold Region Wastewater Lagoon Sludge: Accumulation, Characterization, and Digestion,* Contract Report DACA89-79-C-0011, U.S. Cold Regions Research and Engineering Laboratory, Hanover, N.H., 1981.
29. Sorber, C. A., H. T. Bausum, S. A. Schaub, and M. J. Small: "A Study of Bacterial Aerosols at a Wastewater Irrigation Site," *J. Water Pollution Control Fed.,* 48(10):2367–2379, 1976.
30. Sorber, C. A., and B. P. Sagik: "Indicators and Pathogens in Wastewater Aerosols

and Factors Affecting Survivability," *Wastewater Aerosols and Disease,* EPA 600/9-80-078, U.S. Health Effects Research Laboratory, Cincinnati, Ohio, 1980, pp. 23–35.
31. Sorber, C. A., B. E. Moore, D. E. Johnson, H. J. Hardy, and R. E. Thomas: "Microbiological Aerosols from the Application of Liquid Sludge to Land," *J. Water Pollution Control Fed.,* 56(7):830–836, 1984.
32. U.S. Bureau of Reclamation, U.S. Dept. of the Interior: *Drainage Manual,* U.S. Government Printing Office, Washington, D.C., 1978.
33. U.S. Environmental Protection Agency: *Protection of Public Water Supplies from Ground-Water Contamination,* EPA 625/4-85-016, Center for Environmental Research Information, Cincinnati, Ohio, September 1985.
34. U.S. Environmental Protection Agency: *Process Design Manual Land Treatment of Municipal Wastewater,* EPA 625/1-81-013, EPA Center for Environmental Research Information, Cincinnati, Ohio, 1981.
35. U.S. Environmental Protection Agency: *Process Design Manual Land Treatment of Municipal Wastewater. Supplement on Rapid Infiltration and Overland Flow,* EPA 625/1-81-013a, EPA Center for Environmental Research Information, Cincinnati, Ohio, October 1984.
36. U.S. Environmental Protection Agency: *Estimating Microorganism Densities in Aerosols from Spray Irrigation of Wastewater,* EPA 600/9-82-003, Center for Environmental Research Information, Cincinnati, Ohio, 1982.
37. U.S. Environmental Protection Agency: *Design Manual Municipal Wastewater Stabilization Ponds,* EPA 625/1-83-015, EPA Center for Environmental Research Information, Cincinnati, Ohio, 1983.
38. U.S. Environmental Protection Agency: *Fate of Priority Pollutants in Publicly Owned Treatment Works,* EPA 440/1-82-303, Environmental Protection Agency, Washington, D.C., 1982.
39. Van Schifgaarde, J.: *Drainage for Agriculture,* American Society of Agronomy Series on Agronomy, No. 17, 1974.
40. Wile, I., G. Miller, and S. Black: "Design and Use of Artificial Wetlands," *Ecological Considerations in Wetlands Treatment of Municipal Wastewaters,* Van Nostrand Reinhold, New York, 1985, pp. 26–37.

Chapter

4

Wastewater Stabilization Ponds

Stabilization ponds have been employed for treatment of wastewater for over 3000 years. The first recorded construction of a pond system in the United States was at San Antonio, Texas in 1901. Today, almost 7000 pond systems are used in the United States for the treatment of municipal and industrial wastewaters,[44] under a wide range of weather conditions ranging from tropical to arctic. These pond systems can be used alone or in combination with other wastewater treatment processes.

Wastewater pond systems can be classified by dominant type of biological reaction, duration and frequency of discharge, extent of treatment ahead of the pond, or arrangement among cells (if more than one cell is used). The most basic classification depends on the dominant biological reactions occurring in the pond, the four principal types being:

- Facultative (aerobic-anaerobic) ponds
- Aerated ponds
- Aerobic ponds
- Anaerobic ponds

All four types depend on the interaction of the in situ biological components for treatment and can be considered to be natural treatment systems. General design features and performance expectations are presented in Table 1.1 in Chap. 1.

The most common type is the *facultative pond;* other terms that are commonly applied to this type are *oxidation pond, sewage lagoon,* and *photosynthetic pond.* Facultative ponds are usually 1.2 to 2.5 m (4 to 8 ft) in depth, with an aerobic layer overlying an anaerobic layer, which often contains sludge deposits. The usual detention time is 5 to 30 days. Anaerobic fermentation occurs in the lower layer, and aerobic stabilization occurs in the upper layer. The key to facultative operation is oxygen production by photosynthetic algae and surface reaeration. The oxygen is utilized by the aerobic bacteria in stabilizing the organic material in the upper layer. The algae are necessary for oxygen production, but their presence in the final effluent represents one of the most serious performance problems associated with facultative ponds.

The total containment pond and the controlled discharge pond are forms of facultative ponds. The *total containment pond* is applicable in climates in which the evaporative losses exceed the rainfall. *Controlled discharge ponds* have long detention times, and the effluent is discharged once or twice per year when the effluent quality and stream conditions are satisfactory. A variation of the controlled discharge pond, used in the southern United States, is called a *hydrograph controlled release lagoon.* The pond discharge is matched to periods of high flow in the receiving stream, the stream hydrograph being used as the control.

In an *aerated pond* oxygen is supplied mainly through mechanical, or diffused, aeration. Aerated ponds are generally 2 to 6 m (6 to 20 ft) in depth with detention times of 3 to 10 days. The chief advantage of aerated ponds is that they require less land area. Aerated ponds can be designed as complete mix reactors or as partial mix reactors; in the former case sufficient energy must be used to keep the pond contents in suspension at all times. The basic design of a complete mix reactor is similar to that of an activated sludge system without sludge recycle and is beyond the scope of this book. References 1, 25, 26, and 43 should be consulted.

Aerobic ponds, also called *high-rate aerobic ponds,* maintain dissolved oxygen (DO) throughout their entire depth. They are usually 30 to 45 cm (12 to 18 in) deep, allowing light to penetrate to the full depth. Mixing is often provided to expose all algae to sunlight and to prevent deposition and subsequent anaerobic conditions. Oxygen is provided by photosynthesis and surface reaeration, and aerobic bacteria stabilize the waste. Detention time is short, 3 to 5 days being usual. Aerobic ponds are limited to warm sunny climates and are used infrequently in the United States.

Anaerobic ponds receive such a heavy organic loading that there is no aerobic zone. They are usually 2.5 to 5 m (8 to 16 ft) in depth and have detention times of 20 to 50 days. The principal biological reactions occurring are acid formation and methane fermentation. Anaerobic ponds are usually used for treatment of strong industrial and agricultural wastes or as a pretreatment step where an industry is a significant contributor to a municipal system. They do not have wide application to the treatment of municipal wastewater.

4.1 Preliminary Treatment

In general, the only mechanical or monitoring and control equipment required for wastewater pond systems are flow measurement devices, sampling systems, and pumps. Design criteria and examples for preliminary treatment components can be found in a number of references,[1,25,37,45,47,48] as well as in equipment manufacturers' catalogs. Flow measurement can be accomplished with relatively simple devices, such as Palmer-Bowlus flumes, V-notch weirs, and Parshall flumes used in conjunction with a recording meter. Frequently, flow measurements and 24-h compositing samplers are combined in a common manhole, pipe, or other housing arrangement. If pumping facilities are necessary, the wet well is sometimes used as a point to recycle effluent or to add chemicals for odor control. Pretreatment facilities should be kept to a minimum at pond systems.

4.2 Facultative Ponds

Facultative pond design is based upon removal of the biochemical oxygen demand (BOD); however, the majority of the suspended solids (SS) will be removed in the primary cell of a pond system. Sludge fermentation feedback of organic compounds to the water in a pond system is significant and has an effect on performance. During the spring and fall the thermal overturn of the pond contents can result in resuspension of significant quantities of benthic solids. The rate of sludge accumulation is affected by the liquid temperature, and additional volume is added for sludge accumulation in cold climates. Although SS have a profound influence on performance of pond systems, most design equations simplify incorporation of this influence by using an overall reaction rate constant. Effluent SS generally consist of suspended organism biomass and do not include suspended waste organic matter.

Several empirical and rational models for the design of these ponds have been developed. These include the ideal plug flow and complete mix models, as well as models proposed by Fritz et al.,[7] Gloyna,[9] Larson,[15] Marais,[22] McGarry and Pescod,[24] Oswald et al.,[29] and Thirumurthi.[38] Several produce satisfactory results, but the use of some may be limited by the difficulty in evaluating coefficients or by the complexity of the model.

Area loading rate method

Canter and Englande[4] reported that most states have design criteria for organic loading and/or hydraulic detention time for facultative ponds. These criteria are assumed to ensure satisfactory performance; however, repeated violations of effluent standards by pond systems that meet state design criteria indicate the inadequacy of the criteria. A summary of the state design criteria for each location and actual design values for organic loading and hydraulic detention time for four facultative pond systems evaluated by the Environmental Protection Agency[26,43] are shown in Table 4.1. Also included is a list of the months in which the federal effluent standards for BOD were exceeded. The actual organic loading for the four systems is nearly equal, but the system in Corinne, Utah consistently satisfied the federal effluent standard. This may be a function of the larger number of cells in the Corinne system, seven as compared with three for the others. More hydraulic short-circuiting is likely to occur in the three-cell systems, resulting in an actual detention time that was shorter than in the Corinne system. The detention time may also be affected by the location of the pond cell inlet and outlet structures.

Based on many years of experience, the following loading rates for various climatic conditions are recommended for use in designing facultative pond systems. For average winter air temperatures above 15°C (59°F), a BOD loading rate range of 45 to 90 kg/(ha · day) [40 to 80 lb/(acre · day)] is recommended. When the average winter air temperature ranges between 0° and 15°C (32° and 59°F), the organic loading rate should range between 22 and 45 kg/(ha · day) [20 to 40 lb/(acre · day)]. For average winter temperatures below 0°C (32°F) the organic loading should range from 11 to 22 kg/(ha · day) [10 to 20 lb/(acre · day)].

The BOD loading rate in the first cell is usually limited to 40 kg/(ha · day) [35 lb/(acre · day)] or less, and the total hydraulic detention time in the system is 120 to 180 days in climates in which the average air temperature is below 0°C (32°F). In mild climates in which the air temperature is higher than 15°C (59°F), loadings on the primary cell can be 100 kg/(ha · day) [89 lb/(acre · day)].

TABLE 4.1 Summary of Design and Performance Data from EPA Pond Studies[43]

Location	Organic loading (kg BOD/ha · day)*			Theoretical detention time, days			Months effluent BOD exceeded 30 mg/L
	State design standard	Design	Actual (1974–75)	State design standard	Design	Actual	
Peterborough, N.H.	39.3	19.6	16.2	None	57	107	Oct., Feb., Mar., Apr.
Kilmichael, Miss.	56.2	43.0	17.5	None	79	214	Nov., July
Eudora, Kan.	38.1	38.1	18.8	None	47	231	Mar., Apr., Aug.
Corinne, Utah	45.0*	36.2*	29.7†	180	180	70	None
			14.6‡			88§	

* (kg/ha · day) × 0.889 = lb/(acre · day).
† Primary cell.
‡ Entire system.
§ Estimated from dye study.

Gloyna equation

Gloyna[9] has proposed the following empirical equation for the design of facultative wastewater stabilization ponds:

$$V = (3.5 \times 10^{-5})QL_a\theta^{35-T}ff' \qquad (4.1)$$

where V = pond volume, m^3
Q = influent flow rate, L/day
L_a = ultimate influent BOD or COD, mg/L
θ = temperature correction coefficient
= 1.085
T = pond temperature, °C
f = algal toxicity factor
f' = sulfide oxygen demand

The BOD removal efficiency is projected to be 80 to 90 percent based on unfiltered influent samples and filtered effluent samples. A pond depth of 1.5 m (5 ft) is suggested for systems with significant seasonal variations in temperature and major fluctuations in daily flow. The surface area design using Eq. 4.1 should always be based on a 1 m (3 ft) depth. The algal toxicity factor f is assumed to be equal to 1.0 for domestic wastes and many industrial wastes. The sulfide oxygen demand f' is also equal to 1.0 for sulfate equivalent ion concentration of less than 500 mg/L. The design temperature is usually selected as the average pond temperature in the coldest month. Sunlight is not considered to be critical in pond design but can be incorporated into Eq. 4.1 by multiplying the pond volume by the ratio of sunlight at the design location to the average found in the southwestern United States.

The Gloyna method was evaluated with use of the reference data in Table 4.1. The equation giving the best fit of the data is shown below as Eq. 4.2. There was considerable scatter to the data, but the relationship is statistically significant.

$$V = 0.035Q(\text{BOD})(1.099)^{\text{LIGHT}(35-T)/250} \qquad (4.2)$$

where BOD = five-day BOD (BOD_5) in the system influent, mg/L
LIGHT = solar radiation, langleys
V = pond volume, m^3
Q = influent flow rate, L/day
T = pond temperature, °C

Complete mix model

The Marais and Shaw[23] equation is based on a complete mix model and first-order kinetics. The basic relationship is shown in Eq. 4.3.

$$\frac{C_n}{C_0} = \left[\frac{1}{1 + k_c t_n}\right]^n \qquad (4.3)$$

where C_n = effluent BOD_5 concentration, mg/L
C_0 = influent BOD_5 concentration, mg/L
k_c = complete mix first-order reaction rate, $days^{-1}$
t_n = hydraulic residence time in each cell, days
n = number of equal-sized pond cells in series

The proposed upper limit for the BOD_5 concentration $(C_e)_{max}$ in the primary cells is 55 mg/L to avoid anaerobic conditions and odors. The permissible depth d of the pond, in feet, was found to be related to $(C_e)_{max}$ as follows:

$$(C_e)_{max} = \frac{700}{0.6d + 8} \tag{4.4}$$

where $(C_e)_{max}$ is the maximum effluent BOD, 55 mg/L, and d is the design depth of the pond in feet. (The equation must be used with USCS units for the depth because of the empirical constants, which cannot be converted to metric units.)

The influence of water temperature on the reaction rate is estimated by Eq. 4.5.

$$k_{cT} = k_{c35}(1.085)^{T-35} \tag{4.5}$$

where k_{cT} = reaction rate at water temperature T, day^{-1}
k_{c35} = reaction rate at 35°C
= 1.2 $days^{-1}$
T = operating water temperature, °C

Plug-flow model

The basic equation for the plug-flow model is:

$$\frac{C_e}{C_0} = \exp\,[-k_p t] \tag{4.6}$$

where C_e = effluent BOD_5 concentration, mg/L
C_0 = influent BOD_5 concentration, mg/L
k_p = plug flow first-order reaction rate, $days^{-1}$
t = hydraulic residence time, days

The reaction rate k_p varies with the BOD loading rate as shown in Table 4.2.

The influence of water temperature on the reaction rate constant can be determined with Eq. 4.6*a*.

$$k_{pT} = k_{p20}(1.09)^{T-20} \tag{4.6a}$$

where k_{pT} = reaction rate at temperature T, $days^{-1}$
k_{p20} = reaction rate at 20°C, $days^{-1}$
T = operating water temperature, °C

TABLE 4.2 Variation of the Plug-Flow Reaction Rate Constant with Organic Loading Rate[28]

Organic loading rate, kg/(ha · day)*	k_p,† days^{-1}
22	0.045
45	0.071
67	0.083
90	0.096
112	0.129

* kg/(ha · day) × 0.8907 = lb/(acre · day).
† Reaction rate constant at 20°C.

Wehner-Wilhelm equation

Thirumurthi[38] found that the flow pattern in facultative ponds is somewhere between ideal plug flow and complete mix, and he recommended the use for pond design of the following equation, developed by Wehner and Wilhelm[49] for chemical reactor design.

$$\frac{C_e}{C_0} = \frac{4ae^{1/(2D)}}{(1 + a)^2(e^{a/2D} - (1 - a)^2e^{-a/2D})} \tag{4.7}$$

where C_0 = influent BOD concentration, mg/L
C_e = effluent BOD concentration, mg/L
e = base of natural logarithms = 2.7183
$a = (1 + 4ktD)^2$
k = first-order reaction rate constant, days^{-1}
t = hydraulic residence time, days
D = dimensionless dispersion number
$= H/vL = Ht/L^2$
H = axial dispersion coefficient, area per unit time
v = fluid velocity, length per unit time
L = length of travel path of a typical particle

Thirumurthi[38] prepared the chart shown in Fig. 4.1 to facilitate the use of Eq. 4.7. The dimensionless term kt is plotted versus the percentage of BOD remaining for dispersion numbers ranging from zero for an ideal plug flow unit to infinity for a completely mixed unit. Dispersion numbers measured in wastewater ponds range from 0.1 to 2.0, with most values less than 1.0. The selection of a value for D can dramatically affect the detention time required to produce a given effluent quality. The selection of a design value for k can have an equal effect. If the chart in Fig. 4.1 is not used, Eq. 4.7 can be solved on a trial-and-error basis, as shown in Example 4.1.

Example 4.1 Determine the design detention time in a facultative pond by solv-

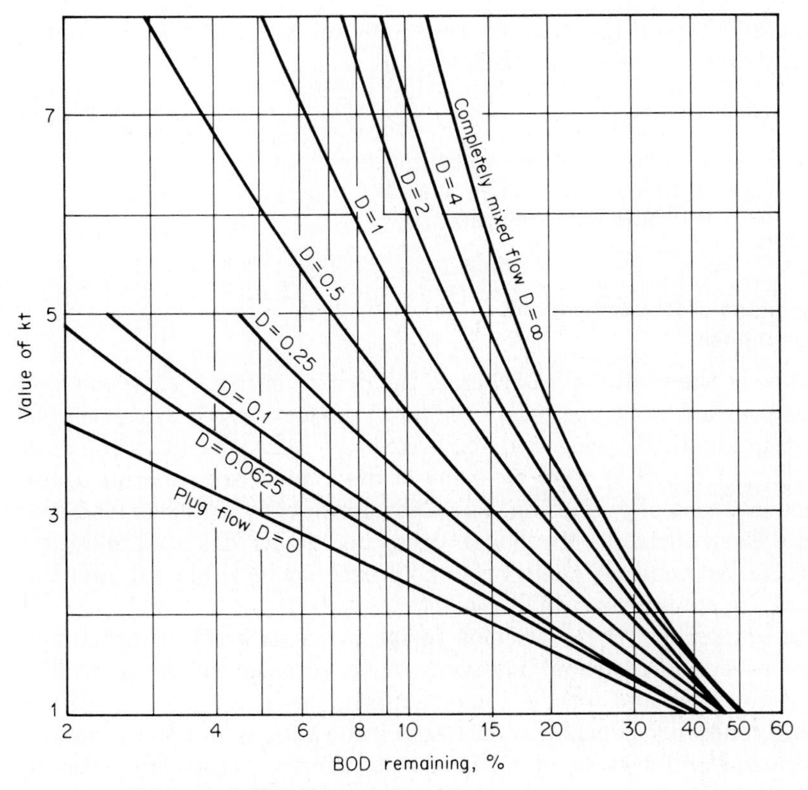

Figure 4.1 Wehner and Wilhelm equation chart.

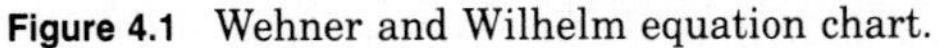

ing Eq. 4.7 on a trial-and-error basis. Assume $C_e = 30$ mg/L, $C_0 = 200$ mg/L, $k_T = 0.028$, and $D = 0.1$.

solution

1. Assume for the first iteration that $t = 50$ days and solve for a:

$$a = (1 + 4k_T Dt)^{1/2}$$
$$= (1 + 4 \times 0.028 \times 0.1 \times 50)^{1/2}$$
$$= 1.25$$

2. Solve Eq. 4.7 and see if the two sides are equal:

$$\frac{C_e}{C_0} = \frac{30}{200} = \frac{(4)(1.25)e^{1/(2)(0.1)}}{(1 + 1.25)^2 e^{1.25/(2)(0.1)} - (1 - 1.25)^2 e^{-1.25/(2)(0.1)}}$$

$$= 0.15 = \frac{742.07}{(5.0625)(518.01) - (0.0625)(0.00193)} = 0.283$$

0.15 does not equal 0.283, so calculation is repeated.

3. The final iteration assumes that $t = 80$ days.

$$0.15 = \frac{817.46}{(5.65)(977.50) - (0.142)(0.00102)} = 0.148$$

The agreement is adequate, so use a design time of 80 days.

The variation of the reaction rate constant k in Eq. 4.7 with water temperature is determined with Eq. 4.8.

$$k_T = k_{20}(1.09)^{T-20} \tag{4.8}$$

where k_T = reaction rate at water temperature T, days^{-1}
k_{20} = reaction rate at 20°C = 0.15 day^{-1}
T = operating water temperature, °C

Comparison of facultative pond design models

Because of the many approaches to the design of facultative ponds it is not possible to recommend the "best" procedure. An evaluation of the design methods presented above with the reference operating data of Table 4.1 failed to show that any of the models are superior to the others in terms of predicting the performance of facultative pond systems.[43] Each of the design models presented above was used to design a facultative pond for the conditions presented in Table 4.3, and the results are summarized in Table 4.4.

The limitations on the various design methods make it difficult to make direct comparisons; however, an examination of the hydraulic detention times and total volume requirements calculated by all the methods show considerable consistency if the Marais and Shaw method is excluded and a value of 1.0 is selected for the dispersion factor in the Wehner-Wilhelm method. The major limitation of all these methods is the selection of a reaction rate constant or other factors in the equations. Even with this limitation, if the pond hydraulic system is designed and constructed, so that the theoretical hydraulic detention time is approached, reasonable success can be ensured with all the design methods. Short circuiting is the greatest deterrent to successful pond

TABLE 4.3 Assumed Conditions for Facultative Design Comparisons

Q = design flow rate = 1893 m^3/day (0.5 million gal/day)
C_0 = influent BOD = 200 mg/L
C_e = required effluent BOD = 30 mg/L
T = water temperature at critical part of year = 10°C
T_a = average winter air temperature = 5°C
Light intensity is adequate.
SS = 250 mg/L
Sulfate = < 500 mg/L

TABLE 4.4 Summary of Results from Facultative Pond Design Methods

Method	Detention time, days		Volume, m^3		Surface area, ha		Primary cell depth, m	Number of cells in series	Organic loading, kg BOD/ha · day	
	Primary cell	Total system	Primary cell	Total system	Primary cell	Total system			Primary	Total
Areal loading rate	53*	71	82,900*	135,300	6.3	11.5	1.7 (1.4)†	4	60	33
Gloyna	—	65	82,900*	123,000	—	12.3	1.5 (1.0)†	—	—	31
Marais & Shaw	17‡	34	32,000‡	64,000	1.3	2.6	2.4	2§	290	145
Plug flow	53*	53	82,900*	123,000	6.3	6.3	1.7 (1.4)†	1§	60	60
Wehner & Wilhelm	53*	36–58	82,900*	68,100 to 109,800	6.3	4.8 to 7.8	1.7 (1.4)†	4	—	80–50

* Controlled by state standards and equal to value calculated for an areal loading rate of 60 kg/(ha · day) and an effective depth of 1.4 m.
† Effective depth.
‡ Also would be controlled by state standards for areal loading rate; however, the method includes a provision for calculating a value, and this calculated value is shown.
§ Baffling recommended to improve hydraulic characteristics.

performance, barring any toxic effects. The importance of the hydraulic design of a pond system cannot be overemphasized.

The surface loading rate approach to design requires a minimum of input data and is based on operational experiences in various geographic areas of the United States. It is probably the most conservative of the design methods, but the hydraulic design still cannot be neglected.

The Gloyna method is applicable only for 80 to 90 percent BOD removal efficiency, and it assumes that solar energy for photosynthesis is above the saturation level. Provision for removals outside this range is not made; however, an adjustment for other solar conditions can be made as described previously. Reference 21 should be consulted if a detailed critique of the Gloyna method is needed.

The Marais and Shaw method is based on complete mix hydraulics, which are not approached in facultative ponds, but the greatest weakness in this approach may lie in the requirement that the primary cell not turn anaerobic. References 20 and 21 can be consulted for a detailed discussion of this model.

Plug-flow hydraulics and first-order reaction kinetics have been found to adequately describe the performance of many facultative pond systems.[26,28,38] A plug-flow model was found to best describe the performance of the four pond systems evaluated in an EPA study.[26,43] Because of the arrangement of most facultative ponds into a series of three or more cells, logically it would be expected that the hydraulic regime could be approximated by a plug-flow model.

Use of the Wehner-Wilhelm equation requires knowledge of both the reaction rate and the dispersion factor, which further complicates the design procedure. If the hydraulic characteristics of a proposed pond configuration are known or can be determined, the Wehner-Wilhelm equation will yield satisfactory results. However, because of the difficulty of selecting both parameters, design with one of the simpler equations is likely to be as good as one using this model.

In summary, all the design methods discussed can provide a valid design if the proper parameters are selected and the hydraulic characteristics of the system are controlled.

4.3 Partial-Mix Aerated Ponds

In the partial-mix aerated pond system, the aeration serves only to provide an adequate oxygen supply, and there is no attempt to keep all the solids in suspension in the pond as with complete-mix and activated-sludge systems. Some mixing obviously occurs and keeps portions of the solids suspended; however, anaerobic degradation of the organic matter that settles does occur. The system is sometimes referred to as a *facultative aerated pond system.*

Even though the pond is only partially mixed, it is conventional to estimate the BOD removal by using a complete-mix model and first-order reaction kinetics. Recent studies[26] have shown that a plug-flow model and first-order kinetics more closely predict the performance of these ponds when either surface or diffused aeration is used. However, most of the ponds evaluated in this study were lightly loaded, and the reaction rates calculated are very conservative because the rate decreases as the organic loading decreases.[28] Because of the lack of better reaction rate design data, it is still necessary to design partial-mix ponds with use of complete mix kinetics.

Partial-mix design model

The design model using first-order kinetics for operation of n equal-sized cells in series is given by Eq. 4.9.

$$\frac{C_n}{C_0} = \frac{1}{[1 + (kt/n)]^n} \tag{4.9}$$

where C_n = effluent BOD concentration in cell n, mg/L
C_0 = influent BOD concentration, mg/L
k = first-order reaction rate constant, days^{-1}
= 0.276 day^{-1} at 20°C (assumed to be constant in all cells)
t = total hydraulic residence time in pond system, days
n = number of cells in the series

If other than a series of equal-volume ponds is to be employed, it is necessary to use the following general equation:

$$\frac{C_n}{C_0} = \left(\frac{1}{1 + k_1t_1}\right)\left(\frac{1}{1 + k_2t_2}\right)\cdots\left(\frac{1}{1 + k_nt_n}\right) \tag{4.10}$$

where $k_1, k_2, \ldots k_n$ are the reaction rates in cells 1 through n (all usually assumed equal for lack of better information) and $t_1, t_2, \ldots t_n$ are the hydraulic residence times in the respective cells.

It has been shown[20] that a number of equal-volume reactors in series is more efficient than reactors of unequal volumes; however, owing to site topography or other factors there may be cases in which it is necessary to construct cells of unequal volume.

Selection of reaction rate constants. The selection of the k value is the critical decision in the design of any pond system. A design value of 0.276 day^{-1} is recommended by the *Ten States Recommended Standards*[37] at 20°C and 0.138 day^{-1} at 1°C. Using these values to calculate the temperature coefficient yields a value of 1.036. Boulier and Atchinson[3] recommend k values of 0.2 to 0.3 at 20°C and 0.1 to 0.15 at 0.5°C. A temperature coefficient of 1.036 results when the two lower or higher

k values are used in the calculation. Reid[33] suggested k values of 0.28 at 20°C and 0.14 at 0.5°C based on research with partial-mix ponds aerated with perforated tubing in central Alaska. These values are essentially identical to the Ten States Standards recommendations.

Influence of number of cells. When using the partial-mix design model, the number of cells in series has a pronounced effect on the size of the pond system required to achieve the specified degree of treatment. The effect can be demonstrated by rearranging Eq. 4.9 and solving for t:

$$t = \frac{n}{k}\left[\left(\frac{C_0}{C_n}\right)^{1/n} - 1\right] \tag{4.11}$$

All terms in this equation have been defined previously.

Example 4.2 Compare detention times for the same BOD removal levels in partial-mix aerated ponds having one to five cells. Assume $C_0 = 200$ mg/L, $k = 0.28\ \text{day}^{-1}$, $T_w = 20°\text{C}$.

solution

1. Solve Eq. 4.11 for a single-cell system:

$$t = \frac{n}{k}\left[\left(\frac{C_0}{C_n}\right)^{1/n} - 1\right]$$
$$= \frac{1}{0.28}\left[\left(\frac{200}{30}\right)^{1/1} - 1\right]$$
$$= 20.2 \text{ days}$$

2. Similarly:

when

$n = 2$ $t = 11$ days
$= 3$ $= 9.4$ days
$= 4$ $= 8.7$ days
$= 5$ $= 8.2$ days

3. Continuing to increase n will result in the detention time approaching the detention time in a plug-flow reactor. It can be seen from the tabulation above that the advantages diminish after the third or fourth cell.

Temperature effects. The influence of temperature on the reaction rate is defined by Eq. 4.12.

$$k_T = k_{20}\theta^{T_w - 20} \tag{4.12}$$

where k_T = reaction rate at temperature T, days^{-1}
k_{20} = reaction rate at 20°C, days^{-1}
θ = temperature coefficient
= 1.036
T_w = temperature of pond water, °C

The pond water temperature (T_w) can be estimated by the following equation, developed by Mancini and Barnhart.[17]

$$T_w = \frac{AfT_a + QT_i}{Af + Q} \tag{4.13}$$

where T_w = pond water temperature, °C (°F)
T_a = ambient air temperature, °C (°F)
T_i = influent water temperature, °C (°F)
A = surface area of pond, m^2 (ft^2)
f = proportionality factor = 0.5
Q = wastewater flow rate, m^3/day (gal/day)

An estimate of the surface area is made on the basis of Eq. 4.11, corrected for temperature, and then the temperature is calculated by Eq. 4.13. After several iterations, when the water temperature used to correct the reaction rate coefficient agrees with the value calculated by Eq. 4.13, the calculation of the detention time in the system is complete.

Pond configuration

The ideal configuration of a pond designed on the basis of complete-mix hydraulics is circular or square; however, even though partial-mix ponds are designed by using the complete-mix model, it is recommended that the cells be designed with a length to width ratio of 3:1 or 4:1. This is done because it is recognized that the hydraulic flow pattern in partial-mix systems more closely resembles the plug-flow condition. The dimensions of the cells can be calculated by Eq. 4.14.

$$V = [LW + (L - 2sd)(W - 2sd) + 4(L - sd)(W - sd)]\frac{d}{6} \tag{4.14}$$

where V = volume of pond or cell, m^3 (ft^3)
L = length of pond or cell at water surface, m (ft)
W = width of pond or cell at water surface, m (ft)
s = slope factor (e.g., 3:1 slope, s = 3)
d = depth of pond, m (ft)

Mixing and aeration

The oxygen requirements control the power input required for partial-mix pond systems. A complete-mix system would require approximately 10 times the power of a system designed to satisfy the oxygen requirements only. There are several rational equations available to estimate the oxygen requirements for pond systems, and these can be found in Refs. 1, 2, 10, and 25. In most partial-mix system design calculations the BOD entering the system is used to estimate the biological oxygen requirements. After calculating the required rate of oxygen transfer, equipment manufacturers' catalogs should be used to determine the zone of complete oxygen dispersion by surface, helical, or air gun aerators or the proper spacing of perforated tubing. Equation 4.15 is used to estimate oxygen transfer rates.

$$N = \frac{N_a}{\alpha \left[\dfrac{C_{sw} - C_L}{C_s} \right] 1.025^{T_w - 20}} \tag{4.15}$$

where N = equivalent oxygen transfer to tap water at standard conditions, kg/h

N_a = oxygen required to treat the wastewater, kg/h (usually taken as 1.5 × the organic loading entering the cell)

α = (oxygen transfer in wastewater)/(oxygen transfer in tap water) = 0.9

C_L = minimum dissolved oxygen concentration to be maintained in the wastewater, assume 2 mg/L

C_s = oxygen saturation value of tap water at 20°C and 1 atm = 9.17 mg/L

T_w = wastewater temperature, °C

C_{sw} = $\beta C_{ss} P$ = oxygen saturation value of the waste, mg/L

β = (wastewater saturation value)/(tap water oxygen saturation value) = 0.9

C_{ss} = tap water oxygen saturation value at temperature T (see Table A.4 in the Appendix for values)

P = ratio of barometric pressure at the pond site to barometric pressure at sea level (assume 1.0 for an elevation of 100 m)

Equation 4.13 can be used to estimate the water temperature in the pond during the summer months, which will be the critical period for design. The use of the partial-mix design procedure is illustrated by Example 4.3.

Example 4.3. Design a four-cell partial-mix aerated pond for the following environmental conditions and wastewater characteristics: Q = 1893 m³/day, C_0 = 200 mg/L, C_e from fourth cell = 30 mg/L, k_{20} = 0.276 day^{-1}, winter air temperature = −5°C, summer air temperature = 30°C, elevation 100 m. Maintain a minimum dissolved oxygen concentration of 2 mg/L in all cells and use a pond depth of 3 m.

solution

1. Assume a winter pond water temperature of 10°C and calculate the volume of a cell in the pond system.

$$k = (0.276)(1.036)^{(10-20)} = 0.194 \text{ day}^{-1}$$

$$t = \frac{4}{0.194}\left[\left(\frac{200}{30}\right)^{1/4} - 1\right] = 12.5 \text{ days}$$

$$t_1 = t_2 = t_3 = t_4 = \frac{12.5}{4} = 3.1 \text{ days}$$

$$V_1 = (3.1)(1893 \text{ m}^3/\text{day}) = 5868 \text{ m}^3$$

2. Assuming that the pond cells have a length to width ratio of 4:1, calculate the dimensions of the cell using Eq. 4.14.

$$V\left(\frac{6}{d}\right) = 4W \times W + (4W - 2 \times 3 \times 3)(W - 2 \times 3 \times 3)$$

$$+ 4(4W - 2 \times 3)(W - 2 \times 3)$$

$$2V = 4W^2 - 90W + 324 + 16W^2 - 120W + 144$$

$$\text{or } W^2 - 8.75W = 0.0833V - 19.5 = 469.5$$

Solve the quadratic equation by completing the square:

$$W^2 - 8.75W + 19.14 = 469.5 + 19.14$$

$$(W - 4.375)^2 = 488.6$$

$$W - 4.375 = 22.10$$

$$W = 26.5 \text{ m}$$

$$L = (26.4)(4) = 106.0 \text{ m}$$

$$\text{Surface area } A = (106.0)(26.5) = 2809 \text{ m}^2$$

3. Check the pond temperature using the calculated cell area of 2809 m^3 and the other known characteristics in Eq. 4.13.

$$T_w = \frac{AfT_a + QT_i}{Af + Q} = \frac{(2809)(0.5)(-5) + (1893)(15)}{(2809)(0.5) + 1893} = 6.5°\text{C}$$

A temperature of 10°C was assumed, so another iteration is necessary.

4. For the second iteration assume 5°C.

$$k = (0.276)(1.036)^{(5-20)} = 0.162 \text{ days}^{-1}$$

Using Eq. 4.11, the total detention time for the four-cell system is 15.0 days, or 3.75 days per cell.

$$V_1 = (3.75)(1893) = 7099 \text{ m}^3$$

$$W^2 - 8.75W = 0.08333V - 19.5$$

$$(W - 4.375) = 32.08$$

$$W = 28.7 \text{ m}$$

$$L = (28.7)(4) = 114.8 \text{ m}$$

$$A = (28.7)(114.8) = 3295 \text{ m}^2$$

$$T_w = \frac{(3295)(0.5)(-5) + (1893)(15)}{(3295)(0.5) + 1893} = 5.7°\text{C}$$

This is close enough to the assumed value of 5°C; therefore, adopt the detention time and cell dimensions calculated in this iteration. Add a freeboard allowance of 0.6 m. This will increase the cell dimensions at the top of the inside of the dike to 40.7 m by 126.8 m. The advantage of a four-cell system was demonstrated in a previous example. In this case using only two cells instead of four will increase the detention time by about 50 perent and increase the surface area and volume by a factor of about 3. This would be undesirable in cold climates because of the enhanced potential for ice formation and in all locations because of the additional costs for construction.

5. Determine the oxygen requirements for this pond system based on the organic loading in each cell and by using Eq. 4.15. The maximum oxygen requirements will occur in the summer months.

Use Eq. 4.13 to estimate pond temperatures.

$$T_w = \frac{(3295)(0.5)(30) + (1893)(15)}{(3295)(0.5 + 1893)} = 22°C$$

At 22°C the tap water oxygen saturation value (C_{ss}) is 8.72 mg/L (see Table A.4 in the Appendix for values).

The organic load in the influent wastewater is:

$$C_0Q = (200 \text{ g/m}^3)(1893 \text{ m}^3\text{/day})(\text{day/24 h})(\text{kg/1000 g})$$
$$= 16 \text{ kg/h}$$

The effluent BOD from the first cell can be calculated by Eq. 4.9.

$$\frac{C_1}{C_0} = \frac{1}{\left[\frac{k_c t}{1} + 1\right]^1} = \frac{1}{(0.162)(3.75) + 1} = 124 \text{ mg/L}$$

Therefore the organic loading on the second cell is

$$\frac{(124 \text{ mg/L})(1893 \text{ m}^3\text{/day})}{(\text{day/24 h})(\text{kg/1000 g})} = 10 \text{ kg/h}$$

Similarly

BOD in cell 2 effluent = 73 mg/L

Organic loading on cell 3 = 6 kg/h

BOD in cell 3 effluent = 42 mg/L

Organic loading on cell 4 = 3 kg/h

The oxygen demand is assumed to be 1.5 times the organic loading, hence

$$N_{a,1} = (1.5)(16 \text{ kg/h}) = 24 \text{ kg/h}$$

Similarly, $N_{a,2} = 15$ kg/h, $N_{a,3} = 9$ kg/h, $N_{a,4} = 4.5$ kg/h. Use Eq. 4.15 to calculate equivalent oxygen transfer.

$$N = \frac{N_a}{\alpha\left(\frac{C_{sw} - C_L}{C_s}\right)1.025^{T_w - 20}}$$

$$C_{sw} = \beta C_{ss} P = (0.9)(8.72 \text{ mg/L})(1.0) = 7.85 \text{ mg/L}$$

$$N_1 = \frac{24 \text{ kg/h of } O_2}{0.9\left(\frac{7.85 - 2.0}{9.17}\right)(1.025)^{22-20}} = 39.7 \text{ kg/h of } O_2$$

Similarly

N_2 = 24.8 kg/h of O_2

N_3 = 14.9 kg/h of O_2

N_4 = 7.4 kg/h of O_2

6. Evaluate both surface and diffused air aeration equipment. A value of 1.9 kg O_2 per kilowatt-hour [1.4 kg/(hp · h)] is recommended for estimating power requirements for surface aerators. A value of 2.7 kg O_2 per kilowatt-hour [2 kg/(hp · h)] is recommended by the manufacturers of diffused aerators. The gas transfer rate must be verified for the equipment selected.

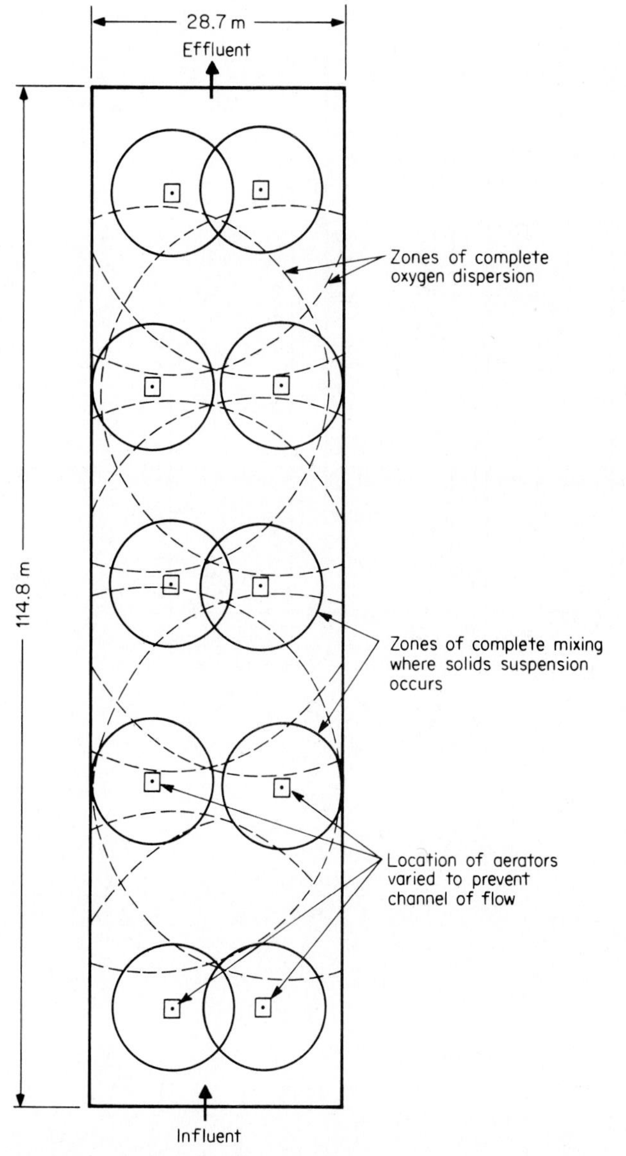

Figure 4.2 Layout of surface aerators in first cell of a partial-mix system.

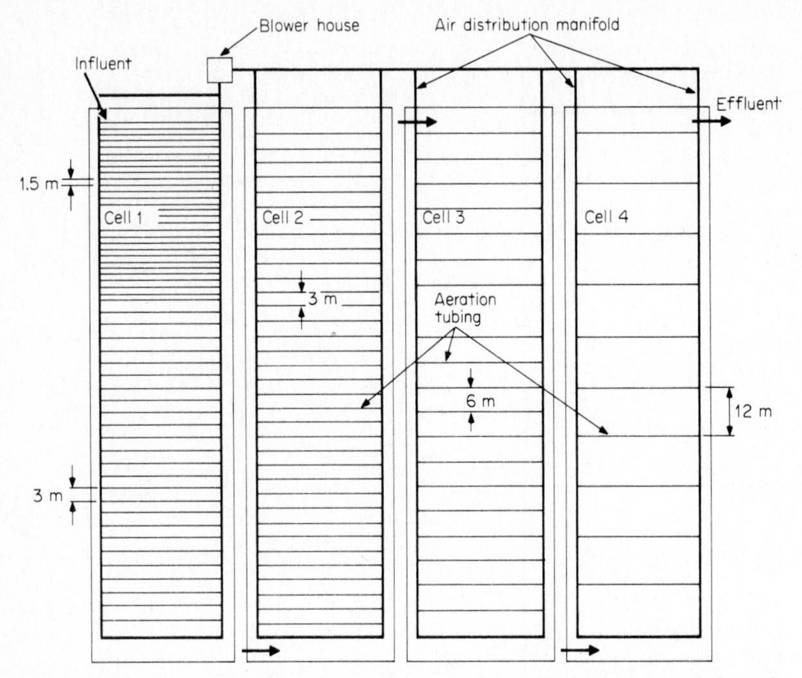

Figure 4.3 Layout of aeration system for partial-mix-diffused-air aerated pond system.

The total power for surface aeration is:

$$\text{Cell 1: } \frac{39.7 \text{ kg/h of } O_2}{1.9 \text{ kg/kWh of } O_2} = 20.9 \text{ kW (28 hp)}$$

Similarly

Cell 2: 13.1 kW (17.5 hp)

Cell 3: 7.8 kW (10.5 hp)

Cell 4: 3.9 kW (5.2 hp)

The total power for diffused aeration is:

$$\text{Cell 1: } \frac{39.7 \text{ kg/h of } O_2}{2.7 \text{ kg/kWh of } O_2} = 14.7 \text{ kW (19.7 hp)}$$

Similarly:

Cell 2: 9.2 kW (12.3 hp)

Cell 3: 5.5 kW (7.4 hp)

Cell 4: 2.7 kW (2.7 hp)

These surface or diffused aerator power requirements must be corrected for gearing and blower efficiency. Assuming 90 percent efficiency for both gearing and blowers, the total power for surface aerators in cell 1 would be 20.9 kW/0.9 =

23.2 kW (31.1 hp). The total power needs are about 51 kW for the surface aerators and 36 kW for the diffused aerators. These are approximate values and are used for the preliminary selection of aeration equipment. The actual power requirement using surface aeration will be determined by using the zone of complete oxygen dispersion reported by the equipment manufacturers, along with the calculated power estimates. The distribution of the two types of aeration equipment are illustrated in Figs. 4.2 and 4.3. Surface aeration equipment is subjected to potential icing problems in cold climates, and use of the fine-bubble perforated tubing requires that a diligent maintenance program be established. A number of communities have experienced clogging of the perforations, particularly in hard water areas. The corrective action required is purging with hydrogen chloride gas.

The final element recommended in this partial-mix aerated pond system is a settling cell with a 2-day detention time.

4.4 Controlled Discharge Ponds

No rational or empirical design model exists specifically for the design of controlled discharge wastewater ponds as utilized in the northern United States and Canada. However, the facultative pond design models may also be applied to the design of controlled discharge ponds provided allowance is made for the required larger storage volumes. The plug-flow model for facultative ponds can be applied to the controlled discharge type if the hydraulic residence time is less than 120 days. In a study of 49 controlled discharge ponds in Michigan, residence times were 120 days or longer and discharge periods ranged from 5 to 30 days for each occurrence. Ponds of this type have been successfully operated in the north central United States by using the following criteria:

- *Overall organic loading:* BOD loading rate 22–28 kg/(ha · day) [20–25 lb/(acre · day)]
- *Liquid depth:* Not more than 2 m (6 ft) in first cell or more than 2.5 m (8 ft) in subsequent cells
- *Hydraulic detention:* At least 6 months storage above the 0.6 m (2 ft) liquid level (including precipitation) but not less than the period of ice cover
- *Number of cells:* At least three for reliability, with interconnected piping for parallel or series operation

The design of the controlled discharge pond must include an analysis showing that receiving stream water quality standards will be maintained during the discharge period and that the receiving watercourses can accommodate the discharge rate from the pond. The design must also develop a recommended discharge schedule.

Selecting the optimum day and hour for release of the pond contents is critical to the success of this method. The operation and maintenance manual must include instructions on how to correlate pond discharge with effluent and stream quality. The pond contents and stream must be carefully examined before and during the disharge period.

The following steps are usually taken for discharge from all systems:

- Isolate the cell to be discharged, usually the final one in series, by valving off the inlet line from the preceding cell.
- Analyze cell contents for parameters of concern in the discharge permit.
- Plan activities to spend full time on control of discharge during the entire discharge period.
- Monitor conditions in receiving stream and request approval from regulatory agency for discharge.
- Commence discharge when approval is received and continue as long as weather is favorable and dissolved oxygen levels and turbidity are below limits. Typically, the last two cells in the series are sequentially isolated and drawn down. Then discharge is interrupted for 1 week or longer while raw wastewater is diverted to one of the cells that has been drawn down. The purpose here is to isolate the first cell prior to its discharge. When the first cell has been drawn down to about 60 cm (24 in) depth, the usual internal series flow pattern without discharge is resumed.
- During the discharge periods samples are taken at least three times daily near the discharge pipe for immediate dissolved oxygen analysis. Additional testing may be required for SS and other parameters.

Experience with the operational concept listed above is limited to northern states with seasonal and climatic constraints on performance. A continuous ice cover on a facultative pond will lower performance, and little better than primary effluent will result if discharge is permitted during such periods. Stringent limits on SS may also limit discharge during the seasonal algal bloom periods. The concept will be quite effective for BOD removal in any location. The process will also work with a more frequent than semiannual discharge cycle, depending on receiving water conditions and requirements.

The hydrograph-controlled release (HCR) pond is a variation of this concept, which was developed for use in the southern United States but can be effectively used in most areas of the country. In this case the discharge periods are controlled by a gauging station in the receiving stream and are allowed to occur during high flow periods. During low flow periods the effluent is stored in the HCR pond. The process design uses conventional facultative or aerated ponds for the basic treatment, followed by the HCR cell for storage/discharge. No treatment allowances are made during design for the residence time in the HCR cell; its sole function is storage. Depending on stream flow conditions, the storage needs may range from 30 to 120 days. The design maximum water level in the HCR cell is typically about 2.4 m (8 ft), with the minimum water level 0.6 m (2 ft). Other physical elements are similar to those used in conventional pond systems. The major advantage of HCR systems is the possibility of utilizing lower discharge standards during high flow conditions as compared with a system designed for very stringent low flow requirements and then operated in that mode on a continuous basis.

4.5 Complete-Retention Ponds

In areas of the world where the moisture deficit, i.e., evaporation minus rainfall, exceeds 75 cm (30 in) annually, a complete-retention wastewater pond may prove to be the most economical method of disposal if low-cost land is available. The pond must be sized to provide the necessary surface area to evaporate the total annual wastewater volume plus the precipitation that would fall on the pond. The system should be designed for the maximum wet year and minimum evaporation year of record if overflow is not permissible under any circumstances. Less stringent design standards may be appropriate in situations in which occasional overflow is acceptable or an alternative disposal area is available under emergency conditions.

Monthly evaporation and precipitation rates must be known to properly size the system. Complete-retention ponds usually require large land areas, and these areas are not productive once they have been committed to this type of system. Land for this system must be naturally flat or be shaped to provide ponds that are uniform in depth and have large surface areas. The design procedure for a complete-retention wastewater pond system is presented in Ref. 43.

4.6 Combined Systems

In certain situations it is desirable to design pond systems in combinations, e.g., an aerated pond followed by a facultative or a tertiary pond. Combinations of this type are designed in essentially the same way as the individual ponds. For example, the aerated pond would be designed as described in Sec. 4.3, and the predicted effluent quality from this unit would be the influent quality for the facultative polishing pond, which would be designed as described in Sec. 4.2. Further details on combined pond systems can be found in Refs. 3, 10, and 34.

4.7 Pathogen Removal

Bacteria, parasite, and virus removal is very effective in multiple-cell wastewater stabilization ponds with suitable detention times. A minimum of three cells is recommended. It is expected that the normal detention time provided for BOD removal in most pond systems may be sufficient to satisfy most regulatory requirements for bacteria and virus removal without additional disinfection; however, a 20-day minimum detention time is suggested. A method for estimating pathogen removal in pond systems is presented in Sec. 3.4.

4.8 Suspended Solids Removal

The occasional high concentration of SS, which can exceed 100 mg/L, in the effluent is the major disadvantage of pond systems. The solids are primarily composed of algae and other pond detritus, not wastewater solids. These high concentrations are usually limited to 2 to 4 months of the year. Several options, discussed in the sections to follow, are available for improving system performance. Further details can be found in Refs. 26, 39, 40, 41, and 43.

Intermittent sand filtration

Intermittent sand filtration is capable of polishing pond effluents at relatively low cost. It is similar to the practice of slow sand filtration in potable water treatment and to the slow sand filtration of raw sewage that was practiced during the early 1900s. Intermittent sand filtration of pond effluents is the application of effluent on a periodic or intermittent basis to a sand filter bed. As the wastewater passes through the bed, SS and other organic matter are removed through a combination of physical straining and biological degradation processes. The particulate matter collects in the top 5 to 8 cm (2 to 3 in) of the filter bed, and this accumulation eventually clogs the surface and prevents effective infiltration of additional effluent. When this happens, the bed is taken out of service, the top layer of clogged sand is removed, and the unit is put back into service. The removed sand can be washed and reused or can be discarded.

The effluent quality is almost totally a function of the gradation of the sand used. When BOD and SS below 30 mg/L will satisfy requirements, a single-stage filter with medium sand will produce a reasonable filter run. If better effluent quality is necessary, a two-stage filtration system should be used, with finer sand in the second stage.

Typical hydraulic loading rates on a single-stage filter range from 0.37 to 0.56 $m^3/(m^2 \cdot day)$ [0.4 to 0.6 million gal/(acre · day)]. If the SS in the influent to the filter will routinely exceed 50 mg/L, the hydraulic loading rate should be reduced to 0.19 to 0.37 $m^3/(m^3 \cdot day)$

[0.2 to 0.4 million gal/(acre · day)] to increase the filter run. In cold weather locations the lower end of the range is recommended during winter operations to avoid the possible need for bed cleaning during the winter months.

The total filter area required for a single-stage operation is obtained by dividing the anticipated influent flow rate by the hydraulic loading rate selected for the system. One spare filter unit should be included to permit continuous operation, since the cleaning operation may require several days. An alternate approach is to provide temporary storage in the pond units. Three filter beds constitute the preferred arrangement to permit maximum flexibility. In small systems that depend on manual cleaning, the individual bed should not be bigger than about 90 m^2 (1000 ft^2). Larger systems with mechanical cleaning equipment might have individual filter beds up to 5000 m^2 (55,000 ft^2) in area.

Selected sands are usually used as the filter media. These are generally described by their *effective size* (e.s.) and *uniformity coefficient* u. The e.s. is the 10 percentile size, i.e., only 10 percent of the filter sand, by weight, is smaller than that size. The uniformity coefficient is the ratio of the 60 percentile size to the 10 percentile size. The sand for single-stage filters should have an e.s. ranging from 0.20 to 0.30 mm and a u of less than 7.0, with less than 1 percent of the sand smaller than 0.1 mm. The u value has little effect on performance, and values ranging from 1.5 to 7.0 are acceptable. In the general case clean, pit-run concrete sand is suitable for use in intermittent sand filters providing the e.s., u, and minimum sand size are suitable.

The design depth of sand in the bed should be at least 45 cm (18 in) plus a sufficient depth for at least 1 year of cleaning cycles. A single cleaning operation may remove 2.5 to 5 cm (1 to 2 in) of sand. A 30-day filter run would then require an additional 30 cm (12 in) of sand. In the typical case an initial bed depth of about 90 cm (36 in) of sand is usually provided. A graded gravel layer 30 to 45 cm (12 to 18 in) thick separates the sand layer from the underdrains. The bottom layer is graded so that its e.s. is four times as great as the openings in the underdrain piping. The successive layers of gravel are progressively finer to prevent intrusion of sand. An alternative is to use gravel around the underdrain piping and then a permeable geotextile membrane to separate the sand from the gravel. Further details on design and performance of these systems can be found in Refs. 26, 35, and 43.

Microstrainers

Early experiments with microstrainers to remove algae from pond effluents were largely unsuccessful. This was generally attributed to the algae being smaller than the mesh size of the microstrainers tested.

A polyester fabric with a 1-μm mesh size has since been developed, and it appears that microstrainers equipped with this fabric are capable of producing an effluent with BOD valves and SS concentrations lower than 30 mg/L.

Microscreen manufacturers are promoting the use of the 1-μm screen with the return of the filtered algae to the pond. Short-term experience indicates that the return of filtered algae does not cause problems; however, the potential exists for the filtered material to accumulate and eventually cause overloading of the screen. The effects of solids recycle through the pond system should be monitored in newly constructed microscreen systems. The first full-scale microstrainer application to pond effluent, a 7200 m^3/day (1.9 million gal/day) unit was placed in operation in Camden, South Carolina in December 1981.[13] Typical design criteria include surface loading rates of 90 to 120 $m^3/m^2 \cdot$ day [1.5 to 2.0 gal/(min $\cdot$ ft^2)] and head losses up to 60 cm (2 ft). Other process variables include drum speed, backwash rate, and pressure; these are normally determined on the basis of influent quality and effluent expectations. The service life of the screen is reported to be about $1\frac{1}{2}$ years, which is considerably less than the manufacturer's prediction of 5 years.

Rock filters

A rock filter operates by allowing pond effluent to travel through a submerged porous rock bed, causing algae to settle out on the rock surfaces as the liquid flows through the void spaces. The accumulated algae are then biologically degraded. Algae removal with rock filters has been studied extensively at Eudora, Kansas; California, Missouri; and Veneta, Oregon.[43]

The principal advantages of the rock filter are its relatively low construction cost and simple operation. Odor problems can occur, and the design life for the filters and the cleaning procedures have not yet been firmly established.

Other solids removal techniques

A detailed discussion of normal granular media filtration, dissolved air flotation, autoflocculation, phase isolation, centrifugation, and coagulation-flocculation is presented in Refs. 26 and 43. These techniques are used infrequently, but the designer should be aware of their potential.

4.9 Nitrogen Removal

The BOD and SS removal capability of pond systems has been reasonably well documented, and reliable designs are possible; however, the nitrogen removal capability of wastewater ponds is given little consideration in most system designs. Nitrogen removal can be critical in many situations, since ammonia nitrogen in low concentrations can adversely affect some young fish in receiving waters. In addition, as described in Chap. 7, nitrogen is often the controlling parameter for design of land treatment systems. Any nitrogen removal in the preliminary pond units can result in very significant savings in the land area required and therefore the costs for land treatment.

Nitrogen loss from streams, lakes, impoundments, and wastewater ponds has been observed for many years, data on nitrogen losses have been insufficient for a comprehensive analysis, and there has been no agreement on the removal mechanisms. Various investigators have suggested algal uptake, sludge deposition, adsorption by bottom soils, nitrification/denitrification, and loss of ammonia as a gas to the atmosphere (volatilization). Recent evaluations[30,32,43] suggest that a combination of factors may be responsible, with the dominant mechanism under favorable conditions being losses to the atmosphere.

The EPA sponsored comprehensive studies of wastewater pond systems in the late 1970s. These results provided absolute verification that significant nitrogen removal does occur in pond systems. Table 4.5 summarizes the key findings from these studies, which confirm that nitrogen removal is in some way related to pH, detention time, and temperature in the pond system. The pH fluctuates as a result of the algae-carbonate interactions in the pond, so wastewater alkalinity is important. Under ideal conditions, up to 95 percent nitrogen removal can be achieved in wastewater stabilization ponds.

Design models

Data were collected on a frequent schedule from every cell in all the pond systems listed in Table 4.5 for at least a full annual cycle. This large body of data allowed quantitative analysis with all major variables included, and two design models were independently developed. These have been validated using the same data from sources not used in the model development. The two models are summarized in Tables 4.6 and 4.7; details on development of Model 1 can be found in Ref. 31.

Both are first-order models, and both depend on pH, temperature, and detention time in the system. Although they both predict the removal of total nitrogen, it is implied in the development of each that volatilization of ammonia is the major pathway for nitrogen removal from wastewater stabilization ponds. Figure 4.4 demonstrates the application of the two models and compares the predicted total nitrogen in the effluent to the actual monthly average values measured at Peterborough, New Hampshire.

Both these models are written in terms of total nitrogen, and they

TABLE 4.5 Data Summary from EPA Pond Studies[31]

Location	Detention time, days	Water temperature, °C	pH (median)	Alkalinity, mg/L	Influent nitrogen, mg/L	Removal, %
Peterborough, N.H., 3 cells	107	11	7.1	85	17.8	43
Kilmichael, Miss., 3 cells	214	18.4	8.2	116	35.9	80
Eudora, Kan., 3 cells	231	14.7	8.4	284	50.8	82
Corinne, Utah, 1st 3 cells	42	10	9.4	555	14.0	46

TABLE 4.6 Design Model Number 1[31]

$$N_e = N_0 \exp\{-k_t[t + 60.6(\text{pH} - 6.6)]\} \tag{4.16}$$

where N_e = effluent total nitrogen, mg/L
N_0 = influent total nitrogen, mg/L
k_T = temperature dependent, rate constant, days^{-1}, pH^{-1}
$= k_{20}\theta^{T-20}$
θ = 1.039
T = water temperature (use Eq. 4.13)

See Refs. 31 and 43 for typical pH values or estimate with:

pH = 7.3exp[0.0005(ALK)]
ALK = expected influent alkalinity, mg/L (derived from data in Refs. 31 and 43)

should not be confused with the still valid equations in Refs. 30 and 43, which are limited to only the ammonia fraction. Calculations and predictions based on total nitrogen should be even more conservative than those earlier models.

The high-rate ammonia removal by air stripping in advanced wastewater treatment depends on high (> 10) chemically adjusted pH. The algae-carbonate interactions in wastewater ponds can elevate the pH to similar levels for brief periods. At other times, at moderate pH levels the rate of nitrogen removal may be low, but the long detention time in the pond compensates.

Application. These models should be useful for new or existing wastewater ponds when nitrogen removal and/or ammonia conversion is required. The design of new systems would typically base detention time on BOD removal requirements. The nitrogen removal that will occur during that time can then be calculated with either model. It is prudent to assume that the remaining nitrogen in the effluent will be ammonia and to then design any further removal/conversion for that amount. If additional land area is available, a final step can be a comparison of the cost of providing additional detention time in the pond for nitrogen removal with the costs for other removal alternatives.

TABLE 4.7 Design Model Number 2[30,43]

$$N_e = N_1 \frac{1}{1 + t(0.000576T - 0.00028)\exp[(1.080 - 0.042T)(\text{pH} - 6.6)]} \tag{4.17}$$

All terms defined in Table 4.6.

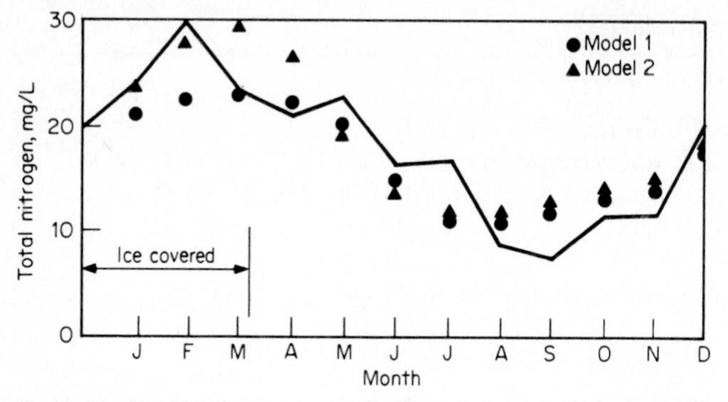

Figure 4.4 Predicted versus actual effluent nitrogen, Peterborough, New Hampshire.

4.10 Phosphorus Removal

The need for phosphorus removal is discussed in Sec. 3.6. In general, phosphorus removal is not often required for wastewaters receiving stabilization pond treatment, but there are a number of exceptions for systems in the north central United States and Canada.

Batch chemical treatment

In order to meet a phosphorus requirement of 1 mg/L for discharge to the Great Lakes, an approach using in-pond chemical treatment in controlled discharge ponds was developed in Canada. Alum, ferric chloride, and lime were all tested by using a motorboat for distribution and mixing of the chemical. A typical alum dosage might be 150 mg/L, and this should produce an effluent from the controlled discharge pond that contains less than 1 mg/L of phosphorus and less than 20 mg/L of BOD and SS. The sludge build-up from the additional chemicals is insignificant and would allow years of operation before requiring cleaning. The costs for this method are very reasonable and much less than for conventional phosphorus removal methods.[11]

Continuous overflow chemical treatment

Studies of in-pond precipitation of phosphorus, BOD, and SS were conducted over a 2-year period in Ontario, Canada.[12] The primary objective of the chemical dosing process was to test removal of phosphorus with ferric chloride, alum, and lime. Ferric chloride doses of 20 mg/L and alum doses of 225 mg/L, when continuously added to the pond influent, effectively maintained pond effluent phosphorus levels below 1 mg/L over a 2-year period. Hydrated lime, at dosages up to 400 mg/L, was not effective in consistently reducing phosphorus below 1 mg/L (1 to 3 mg/L was achieved) and produced no BOD reduction, while slightly increasing the SS concentration. Ferric chloride reduced effluent BOD from 17 to 11 mg/L and SS from 28 to 21 mg/L; alum produced no BOD reduction and a slight SS reduction (from 43 to 28 to 34 mg/L). Consequently, direct chemical addition appears to be effective only for phosphorus removal.

A six-cell pond system located in Waldorf, Maryland was modified to operate as two three-cell units in parallel.[5] One system was used as a control, and alum was added to the other for phosphorus removal. Each system contained an aerated first cell. Alum addition to the third cell of the system proved to be more efficient in removing total phosphorus, BOD, and SS than alum addition to the first cell: total phosphorus reduction averaged 81 percent when alum was added to the inlet to the third cell and 60 percent when alum was added to the inlet of the first cell. Total phosphorus removal in the control ponds averaged 37 percent. When alum was added to the third cell, the effluent total phosphorus concentration averaged 2.5 mg/L, with the control units averaging 8.3 mg/L. Improvements in BOD and SS removal by alum addition were more difficult to detect, and at times increases in effluent concentrations were observed.

4.11 Physical Design and Construction

Regardless of the care taken to evaluate coefficients and apply biological or kinetic models, if sufficient consideration is not given to optimization of the pond layout and construction, the actual efficiency may be far less than the calculated efficiency. The physical design of a wastewater pond is as important as the biological and kinetic design. The biological factors affecting wastewater pond performance are primarily employed to estimate the required hydraulic residence time to achieve a specified efficiency. Physical factors, such as length to width ratio, will determine the actual efficiency achieved.

Length to width ratios are determined according to the design model used. Complete-mix ponds should have a length to width ratio of 1:1, whereas plug-flow ponds require a ratio of 3:1 or greater.

The danger of groundwater contamination may impose seepage restrictions, necessitating lining or sealing of the pond. Reuse of pond effluents in dry areas, where all water losses are to be avoided, may also dictate the use of linings. Layout and construction criteria should be established to reduce dike erosion from wave action, weather, rodent attacks, etc. Transfer structure placement and size affect flow patterns within the pond and determine operational capabilities in controlling the water level and discharge rate.

Dike construction

Dike stability is most often affected by erosion caused by wind-driven wave action or by rain and rain-induced weathering. Dikes may also be destroyed by burrowing rodents. A good design will anticipate these problems and provide a system that can, through cost-effective operation and maintenance, keep all three under control.

Erosion protection is necessary on all slopes; however, if winds are predominantly from one direction, protection should be emphasized for those areas that receive the full force of the wind-driven waves. Protection should extend from at least 0.3 m (1 ft) below the minimum water level to at least 0.3 m above the maximum water surface. Asphalt, concrete, fabric, low grasses, and riprap have all been used to provide protection from wave action. The use of riprap can however, make weed and rodent control more difficult. In some cases when fabric liners are used, a covering of riprap is also used to protect the plastic materials from damaging ultraviolet radiation from the sun. Rodent control can be achieved with earthen dikes by periodically changing the water levels to flood the burrows. The selection of proper soils and compaction during construction can render an earthen dike essentially impermeable. Seepage collars should be provided around any pipe penetrating the dike; these collars should extend a minimum of 0.6 m (2 ft) from the pipe.

Pond sealing

The primary motive for sealing ponds is to prevent seepage, which can pollute groundwaters and affect treatment performance by causing fluctuations in water depth. Sealing methods can be grouped in three categories:

- Synthetic and rubber liners
- Compacted earth or soil-cement liners
- Natural and chemical treatment liners

Within each category there is a wide variety of application characteristics. Choosing the appropriate lining for a specific site is a critical factor in pond design and seepage control. Seepage rates range from 0.003 cm/day (0.001 in/day) for synthetic membranes to about 10 cm/day (4 in/day) for soil-cement liners.[43] Detailed information is available from manufacturers and in other publications.[14,27]

Pond hydraulics

In the past the majority of ponds were designed to receive influent wastewater through a single pipe, usually located toward the center of the first cell in the system. However, hydraulic and performance studies[6,8,18,19] have shown that a center discharge point is not the most efficient point at which to introduce wastewater to a pond. Multiple inlet arrangements are preferred even in small ponds [< 0.5 ha (< 1.2 acre)]. The inlet points should be as far apart as possible, and the water should preferably be introduced by means of a long diffuser. The inlets and outlets should be placed so that the velocity profile of flow through the pond is uniform between successive inlets and outlets.

Single inlets can be used successfully if the inlet is located at the greatest possible distance from the outlet structure and is baffled, or if the flow is otherwise directed to avoid currents and short circuiting. Outlet structures should be designed for multiple-depth withdrawal, and all withdrawals should be at a minimum of 0.3 m (1 ft) below the water surface to reduce the potential impact of algae and other surface detritus on effluent quality.

Analysis of performance data from selected aerated and facultative ponds indicates that four cells in series are desirable to give the best BOD and fecal coliform removals for ponds designed as plug-flow systems. Good performance can also be obtained with a smaller number of cells if baffles or dikes are used to optimize the hydraulic characteristics of the system.

Better treatment is obtained when the flow is guided more carefully through the pond. In addition to treatment efficiency, economics and aesthetics play an important role in deciding whether or not baffling is desirable. In general, the more baffling that is used, the better the flow control and treatment efficiency. The lateral spacing and length of the baffle should be specified so that the cross-sectional area of flow is as close to a constant as possible.

Wind generates a circulatory flow in bodies of water. To minimize short circuiting due to wind, the pond inlet-outlet axis should be aligned perpendicular to the prevailing wind direction if possible. If this is not possible, baffling can be used to control to some extent the wind-induced circulation. In a constant-depth pond the surface current will be in the direction of the wind, and the return flow will be in the upwind direction along the bottom.

Ponds that are stratified because of temperature differences between the inflow and the pond contents tend to behave differently in winter and summer. In summer the inflow is generally colder than the pond, so it sinks to the pond bottom and flows toward the outlet. In the winter the reverse is generally true, and the inflow rises to the surface and flows toward the outlet. A likely consequence is that the effective treatment volume of the pond is reduced to that of the stratified inflow layer (density current). The result can be a drastic decrease in detention time and an unacceptable level of treatment.

4.12 Storage Ponds for Land Treatment Systems

Ponds for seasonal effluent storage are sometimes required in conjunction with the land treatment systems described in Chap. 7. Storage is necessary for all nonoperational periods in the land treatment system and is desirable for flow equalization and emergency system backup. Nonoperating periods may be due to climate, planting, or harvesting and other maintenance operations. The design storage volume is determined from a calculated water balance during design, as described in Chap. 7 (see Example 7.3 for the procedure).

The storage pond may follow other conventional treatment units or may be the final cell in a stabilization pond system. The storage cell is usually deeper than typical treatment pond cells and can range from 3 to 6 m (10 to 20 ft) in depth. Credit should be taken during design for the additional treatment that will occur in this storage pond by using the methods presented in this chapter and in Chap. 3. Calculation of nitrogen removal by either Eq. 4.16 or Eq. 4.17 is particularly important. Nitrogen is often the limiting design factor for land treatment systems, directly affecting the land area required for treatment. Any nitrogen removal in the storage pond will reduce the final treatment area and the costs. Similarly, the pathogen removal in the pond can often satisfy requirements without further disinfection.

The operation of the storage pond will depend on the type of land treatment system in use. Storage is usually only provided for emergencies in rapid infiltration systems, so the pond is drained as soon as it is possible to do so. Since overland flow systems are not very effective for algae removal (see Chap. 7 for details), storage ponds for these systems are bypassed during algal bloom periods, and the ponds are drawn down when algae concentrations are low. Algae are not a concern for slow rate land treatment, so the storage pond may stay on line continuously. This is necessary if nitrogen or pathogen removal is expected in the storage cell. In this case, treated wastewater flow into the cell should continue on a year-round basis, and withdrawals should be scheduled for attainment of the specified water depth at the end of the operating season for the land treatment component.

REFERENCES

1. Al-Layla, M. A., S. Ahmad, and E. J. Middlebrooks: *Handbook of Wastewater Collection and Treatment: Principles and Practices,* Garland STPM Press, New York, 1980.
2. Benefield, L. D., and C. W. Randall: *Biological Process Design for Wastewater Treatment,* Prentice-Hall, Englewood Cliffs, N.J., 1980.
3. Boulier, G. A., and T. J. Atchinson: "Practical Design and Application of the Aerated-Facultative Lagoon Process," Hinde Engineering Co., Highland Park, Ill., 1975.

4. Canter, L. W., and A. J. Englande: "States' Design Criteria for Waste Stabilization Ponds," *J. Water Pollution Control Fed.*, 42(10):1840–1847, 1970.
5. Engel, W. T., and T. T. Schwing: *Field Study of Nutrient Control in a Multicell Lagoon*, EPA 600/2-80-155, Environmental Protection Agency, Municipal Engineering Research Laboratory, Cincinnati, Ohio, 1980.
6. Finney, B. A., and E. J. Middlebrooks: "Facultative Waste Stabilization Pond Design," *J. Water Pollution Control Fed.*, 52(1):134–147, 1980.
7. Fritz, J. J., A. C. Middleton, and D. D. Meredith: "Dynamic Process Modeling of Wastewater Stabilization Ponds," *J. Water Pollution Control Fed.*, 51(11):2724–2743, 1979.
8. George, R. L.: "Two-Dimensional Wind-Generated Flow Patterns, Diffusion and Mixing in a Shallow Stratified Pond," Ph.D. dissertation, Utah State University, Logan, 1973.
9. Gloyna, E. F.: "Facultative Waste Stabilization Pond Design," *Ponds as a Waste Treatment Alternative, Water Resources Symp. no. 9*, University of Texas, Austin, 1976.
10. Gloyna, E. F.: *Waste Stabilization Ponds*, Monograph Series No. 60, World Health Organization, Geneva, 1971.
11. Graham, H. J., and R. B. Hunsinger: "Phosphorus Removal in Seasonal Retention by Batch Chemical Precipitation," Project No. 71-1-13, Wastewater Technology Centre, Environment Canada, Burlington, Ontario (undated).
12. Graham, H. J., and R. B. Hunsinger: "Phosphorus Reduction from Continuous Overflow Lagoons by Addition of Coagulants to Influent Sewage," Res. Rep. No. 65, Ontario Ministry of the Environment, Toronto, 1977.
13. Harrelson, M. E., and J. B. Cravens: "Use of Microscreens to Polish Lagoon Effluent," *J. Water Pollution Control Fed.*, 54(1):36–42, 1982.
14. Kays, W. B.: *Construction of Linings for Reservoirs, Tanks, and Pollution Control Facilities*, 2d ed., Wiley-Interscience, John Wiley, New York, 1986.
15. Larson, T. B.: "A Dimensionless Design Equation for Sewage Lagoons," Dissertation, University of New Mexico, Albuquerque, 1974.
16. Malina, J. F., R. Kayser, W. W. Eckenfelder, Jr., E. F. Gloyna, and W. R. Drynan: "Design Guides for Biological Wastewater Treatment Processes," Report CRWR-76, Center for Research in Water Resources, University of Texas, Austin, 1972.
17. Mancini, J. L., and E. L. Barnhart: "Industrial Waste Treatment in Aerated Lagoons," *Ponds as a Wastewater Treatment Alternative, Water Resources Symp. No. 9*, University of Texas, Austin, 1976.
18. Mangelson, K. A., and G. Z. Watters: "Treatment Efficiency of Waste Stabilization Ponds," *J. Sanit. Eng. Div.*, ASCE, 98(SA2), 1972.
19. Mangelson, K. A.: Hydraulics of Waste Stabilization Ponds and Its Influence on Treatment Efficiency," Ph.D. dissertation, Utah State University, Logan, 1971.
20. Mara, D. D.: *Sewage Treatment in Hot Climates*, John Wiley, New York, 1976.
21. Mara, D. D.: "Discussion," *Water Res.*, 9:595, 1975.
22. Marais, G. V. R.: "Dynamic Behavior of Oxidation Ponds," *Proceedings of Second International Symposium for Waste Treatment Lagoons*, Kansas City, Mo., June 23–25, 1970.
23. Marais, G. V. R., and V. A. Shaw: "A Rational Theory for the Design of Sewage Stabilization Ponds in Central and South Africa," *Trans. S. Afr. Inst. Civil Engrs*, 3:205, 1961.
24. McGarry, M. C., and M. B. Pescod: "Stabilization Pond Design Criteria for Tropical Asia," *Proceedings of Second International Symposium for Waste Treatment Lagoons*, Kansas City, Mo., June 23–25, 1970.
25. Metcalf and Eddy: *Wastewater Engineering Treatment Disposal Reuse*, McGraw-Hill, New York, 1979.
26. Middlebrooks, E. J., C. H. Middlebrooks, J. H. Reynolds, G. Z. Watters, S. C. Reed, and D. B. George: *Wastewater Stabilization Lagoon Design, Performance, and Upgrading*, Macmillan, New York, 1982.
27. Middlebrooks, E. J., C. D. Perman, and I. S. Dunn: *Wastewater Stabilization Pond Linings*, Special Report 78-28, Cold Regions Research and Engineering Laboratory, Hanover, N.H., 1978.

28. Neel, J. K., J. H. McDermott, and C. A. Monday: "Experimental Lagooning of Raw Sewage," *J. Water Pollution Control Fed.,* 33(6):603–641, 1961.
29. Oswald, W. J., A. Meron, and M. D. Zabat: "Designing Waste Ponds to Meet Water Quality Criteria," *Proceedings of Second International Symposium for Waste Treatment Lagoons,* Kansas City, Mo., June 23–25, 1970.
30. Pano, A., and E. J. Middlebrooks: "Ammonia Nitrogen Removal in Facultative Wastewater Stabilization Ponds," *J. Water Pollution Control Fed.,* 54(4):344–351, 1982.
31. Reed, S. C.: *Nitrogen Removal in Wastewater Ponds,* CRREL Report 84-13, Cold Regions Research and Engineering Laboratory, Hanover, N.H., June 1984.
32. Reed, S. C.: *Wastewater Stabilization Ponds: An Update on Pathogen Removal,* Environmental Protection Agency, Office of Municipal Pollution Control, Washington, D.C., August 1985.
33. Reid, L. D., Jr.: Design and Operation for Aerated Lagoons in the Arctic and Subarctic," Report 120, U.S. Public Health Service, Arctic Health Research Center, College, Alaska, 1970.
34. Rich, L. G.: "Design Approach to Dual-Power Aerated Lagoons," *J. Environ. Eng. Div.,* ASCE, 108(EE3):532, 1982.
35. Russell, J. S., E. J. Middlebrooks, and J. H. Reynolds: *Wastewater Stabilization Lagoon-Intermittent Sand Filter Systems,* EPA 600/2-80-032, Environmental Protection Agency, Municipal Engineering Research Laboratory, Cincinnati, Ohio, 1980.
36. Swanson, G. R., and K. J. Williamson: "Upgrading Lagoon Effluents with Rock Filters," *J. Environ. Eng. Div.,* ASCE, 106(EE6):1111–1119, 1980.
37. *Ten States Recommended Standards for Sewage Works.* A Report of the Committee of Great Lakes–Upper Mississippi River Board of State Sanitary Engineers, Health Education Services Inc., Albany, N.Y., 1978.
38. Thirumurthi, D.: "Design Criteria for Waste Stabilization Ponds," *J. Water Pollution Control Fed.,* **46**(9):2094–2106, 1974.
39. U.S. Environmental Protection Agency: *Process Design Manual for Suspended Solids Removal,* EPA 625/1-75-003a, Center for Environmental Research Information, Cincinnati, October 1975.
40. U.S. Environmental Protection Agency: *Upgrading Lagoons,* Technology Transfer Document, Environmental Protection Agency, Washington, August 1973.
41. U.S. Environmental Protection Agency: *Process Design Manual for Upgrading Existing Wastewater Treatment Plants, Technology Transfer,* Environmental Protection Agency, Washington, October 1974.
42. U.S. Environmental Protection Agency: *Process Design Manual for Land Treatment of Municipal Wastewater,* EPA 625/1-81-013, Center for Environmental Research Information, Cincinnati, 1981.
43. U.S. Environmental Protection Agency: Design Manual: *Municipal Wastewater Stabilization Ponds,* EPA 625/1-83-015, Center for Environmental Research Information, Cincinnati, 1983.
44. U.S. Environmental Protection Agency: *The 1980 Needs Survey,* EPA 430/9-81-008, Office of Water Program Operations, Washington, 1981.
45. U.S. Environmental Protection Agency: *Design Criteria for Mechanical, Electrical and Fluid System and Component Reliability,* EPA 430/99-74-001, Office of Water Program Operations, Washington, 1974.
46. Wallace, A. T.: "Land Application of Lagoon Effluents," in: *Performance and Upgrading of Wastewater Stabilization Ponds,* EPA 600/9-79-011, Environmental Protection Agency Municipal Engineering Research Laboratory, Cincinnati, Ohio, 1978.
47. Water Pollution Control Federation and American Society of Civil Engineers: *Wastewater Treatment Plant Design,* MOP/8, Water Pollution Control Federation, Washington, D.C., 1977.
48. Water Pollution Control Federation: *Preliminary Treatment for Wastewater Facilities,* MOP/OM-2, Washington, D.C., 1980.
49. Wehner, J. F., and R. H. Wilhelm: "Boundary Conditions of Flow Reactor," *Chem. Eng. Sci.,* 6:89–93, 1956.

Chapter

5

Aquaculture Systems

Aquaculture is defined as the use of aquatic plants or animals as a component in a wastewater treatment system. In many parts of the world wastewater is used for the production of fish or other forms of aquatic biomass in aquaculture operations. Some degree of wastewater renovation may occur in these cases but it is not the primary intent. The major focus in this chapter is on those systems where wastewater treatment is the functional intent of the operation.

Aquaculture treatment systems can utilize one major type of plant or animal in a monoculture operation, or use a variety of plants and animals in a polyculture operation. Both marine (seawater) and freshwater concepts have been tested. The major biological components include: floating plants, fish and other animals, planktonic organisms, and submerged plants. Emergent plants are also used, but these are more characteristic of wetland systems and are discussed in Chap. 6.

The treatment responses in an aquaculture system are due either to the direct uptake of material by the plants or animals and by the presence of these biota altering the physical environment in the system, or, as in the case of water hyacinths, the plant roots' acting as the host substrate for attached microbial organisms which provide a very significant degree of treatment. All of these plants and animals have specific environmental requirements that must be maintained for their successful use, and in most cases a regular harvest is necessary to ensure optimum performance. Performance expectations for these systems are listed in Table 1.1 in Chap. 1 and elsewhere in this chapter.

5.1 Floating Plants

Aquatic plants have the same basic nutritional requirements as plants growing on land and are influenced by many of the same environmental factors. The floating aquatic plants with the greatest known potential for wastewater treatment include water hyacinths, duckweeds, pennywort, and water ferns. Table 5.1 provides information on distribution of these plants in the United States and some of the critical environmental requirements. Hyacinths, pennywort, and duckweeds are the only varieties tested to date with wastewater, in pilot or full-scale systems.

Water hyacinths

Water hyacinth (*eichhornia crassipes*) is a perennial, freshwater aquatic macrophyte (water tolerant vascular plant) with rounded, upright, shiny green leaves and spikes of lavender flowers. The morphology of a typical hyacinth plant is shown in Fig. 5.1.

The petioles of the plant are spongy with many air spaces and contribute to the buoyancy of the hyacinth plant. The size varies with habitat. The root length will vary with the nutrient status of the water and the frequency of plant harvest. In nutrient-rich wastewaters with regular harvests the roots might extend 10 cm (4 in) below the central rhizome. If harvests are not performed the roots can grow and penetrate the substrate in unlined basins. The plant will also grow in moist soils. When grown in wastewater individual plants range from 50 to 120 cm (20 to 47 in) from the top of the flower to the root tips.

The hyacinth flower produces seeds but the principal means of reproduction is via offshoots (stolons) from the underwater rhizome, as shown on Fig. 5.1, which result in an interconnected, dense mat of plants on the water surface. The plants spread laterally until the water surface is covered and then the vertical growth increases. Hyacinths are one of the most productive photosynthetic plants in the world. It has been estimated that 10 plants could produce 600,000 more during an 8 month growing season and completely cover 0.4 ha (1 acre) of a natural freshwater surface.[19] The rate can be even higher in wastewater ponds. Wolverton has estimated a productivity of 140 metric tons/(ha · year) [154 ton/(acre · year)] for hyacinths grown in wastewater ponds.[32] This very rapid growth is the reason that hyacinths are a serious nuisance problem in southern waterways, but these same attributes become an advantage when used in a wastewater treatment

TABLE 5.1 Floating Aquatic Plants for Wastewater Treatment[25]

Common name, *Scientific name*	Distribution	Temperature, °C		Maximum salinity tolerance, mg/L	Optimum pH
		Desirable	Survival		
Water hyacinth		20–30	10	800	5–7
Eichhornia crassipes	Southern U.S.				
Water fern		>10	5	2500	3.5–7
Azolia caroliniana	Throughout U.S.				
Azolia filculoides	Throughout U.S.				
Duckweed		20–30	5	3500	5–7
Spirodela plyrihiza	Throughout U.S.				
Lemna trisculca	Northern U.S.				
Lemna obscura	Eastern and southern U.S.				
Lemna minor	Throughout U.S.				
Lemna gibba	Great Plains and western U.S.				
Wolfia spp.	Throughout U.S.				

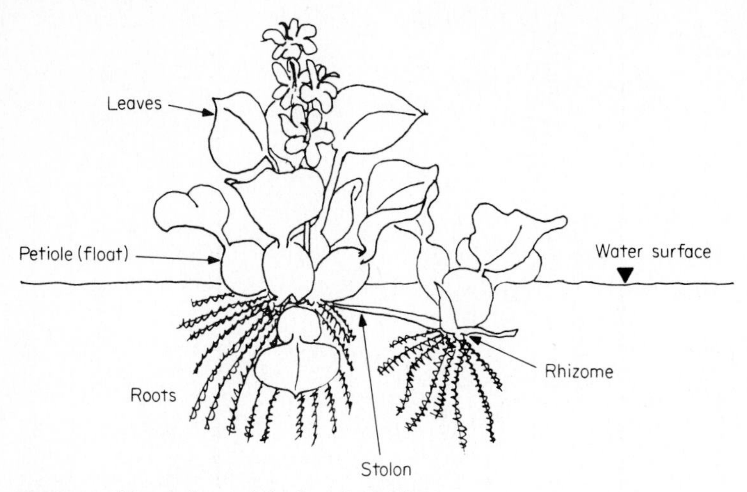

Figure 5.1 Morphology of the hyacinth plant.

system. Because of its history as a nuisance weed the interstate transport of hyacinth plants is prohibited by federal law.

The roots, petioles, flower stalks, and stolons all originate at the basal rhizome. Under freezing conditions the leaves and flowers will die and may expose the upper tip of the rhizome. The plant can regenerate from an undamaged rhizome but if the tip freezes the entire plant will die. This sensitivity to low temperatures is the major factor limiting the natural range of the hyacinth plant and its use in unprotected wastewater treatment facilities. Fig. 2.1 defines the areas suitable for unprotected wastewater treatment systems using hyacinths. Short-term summer use might be possible further north than shown on Fig. 2.1 but this would require a greenhouse for culturing and protecting the plants during the remainder of the year. The protective shelters and heat that would be required to sustain a year-round operation north of the zones shown on Fig. 2.1 are probably not cost effective.[16]

The dry weight composition of water hyacinth plants removed from wastewater systems is given in Table 5.2. The major constituent of the bulk plant is water, comprising about 95 percent of the total mass. This very high water content is a significant factor in the economics of the various disposal/utilization options for the harvested plants.

Performance expectations. Water hyacinth systems are capable of removing high levels of biochemical oxygen demand (BOD), suspended solids (SS), metals, and nitrogen, and significant levels of trace organics. The treatment concept has been developed through extensive laboratory and pilot scale research as well as evaluation of full-scale facilities.

TABLE 5.2 Composition of Hyacinth Plants Grown in Wastewater

Constituent	Percent of dry weight	
	Average	Range
Crude protein	18.1	9.7–23.4
Fat	1.9	1.6–2.2
Fiber	18.6	17.1–19.5
Ash	16.6	11.1–20.4
Carbohydrate	44.8	36.9–51.6
Kjeldahl nitrogen (as N)	2.9	1.6–3.7
Phosphorus (as P)	0.6	0.3–0.9

Hyacinths can be used to upgrade existing systems or to produce secondary, advanced secondary, or tertiary effluents depending on the design loading rates and management practices used.

Hyacinths on the water surface of a pond create a totally different environmental condition in the water as compared to an exposed water surface. The dense canopy of leaves shades the surface and prevents algal growth. This in turn maintains the liquid pH at near neutral levels. The mass of plants on the surface also minimizes wind-induced turbulence and mixing, as well as surface reaeration, and moderates water temperature fluctuations. As a result, the near surface water tends to be low in oxygen and the benthic zone is usually anaerobic even in shallow ponds.

The plant can survive and grow in anaerobic waters since oxygen is transmitted from the leaves to the root mass. The attached biological growth on the root mass is similar to trickling filter and RBC slimes, but in this case the oxygen source (from the roots) is near the center of the mass rather than on the outside. Bacteria, fungi, predators, filter feeders, and detritovores have been reported in large numbers on and among the plant roots. Typical performance data from several systems are given in Table 5.3. The excellent performance of the Coral Springs, Florida system is believed to be in part due to the use of multiple cells and the shallow 38 cm (15 in) depth, which allows a greater portion of the contained wastewater to contact the root zone of the plants.

BOD removal. The removal of BOD in a hyacinth pond is caused by the same factors described in Chap. 4 for conventional stabilization ponds. Further, very significant treatment contributions can be ascribed to the attached growth on the plant roots. The efficiency of BOD removal will be directly related to the density of the plant cover and the depth of water in the system. At water depths of 1 to 2 m (3 to 6 ft) a BOD loading of about 6.7×10^{-4} kg/kg of wet plant mass per day has been recommended by Wolverton[31] when facultative pond ef-

TABLE 5.3 Performance of Hyacinth Wastewater Treatment Systems

Location	BOD, mg/L		SS, mg/L		Total N, mg/L		Total P, mg/L		Reference
	Influent	Effluent	Influent	Effluent	Influent	Effluent	Influent	Effluent	
National Space Tech. Lab., Miss.*	110	7	97	10	12	3.4	3.7	1.6	31, 32
Lucedale, Miss.†	161	23	125	6	—	—	—	—	33
Orange Grove, Miss.‡	50	14	49	15	—	—	—	—	33
Williamson Cr., Tex.§	46	6	91	8	7.7	3.3	7	5.7	6
Coral Springs, Fla.¶	13	3	—	3	22.4	1.0	11	3.6	23

* Single, facultative cell, 122 cm deep, 2 ha, detention time 54 days, hydraulic loading 240 $m^3/(ha \cdot day)$, organic load 26 $kg/(ha \cdot day)$ BOD.
† Single, facultative cell, 173 cm deep, 3.6 ha, detention time 67 days, hydraulic loading 260 $m^3/(ha \cdot day)$, organic load 44 $kg/(ha \cdot day)$ BOD.
‡ Two-cell aerated pond, 183 cm deep, 0.3 ha, detention time 7 days, hydraulic loading 3570 $m^3/(ha \cdot day)$, organic loading 179 $kg/(ha \cdot day)$ BOD.
§ Four-cell facultative pond, 85 cm deep, 0.06 ha, detention time 4.5 days, hydraulic loading 109 $m^3/(ha \cdot day)$, organic loading 89 $kg/(ha \cdot day)$ BOD.
¶ Five-cell facultative pond, 38 cm deep, 0.5 ha, detention time 11 days, hydraulic loading 378 $m^3/(ha \cdot day)$, organic loading 113 $kg/(ha \cdot day)$ BOD.

fluent is applied to the hyacinth cells. Assuming 100 percent coverage of dense plants on the water surface this translates to a surface loading of about 225 kg/(ha · day) [200 lb/(acre · day)] BOD. At 80 percent surface coverage a loading of 140 kg/(ha · day) BOD has been recommended by Wolverton.[32]

Suspended solids removal. The removal of suspended solids occurs through entrapment in the plant root zone and by gravity sedimentation in the quiescent water beneath the surface mat of hyacinth plants. Due to the less turbulent water conditions, sedimentation will be more effective in a hyacinth pond than in a conventional pond with an open water surface. Another major contribution to solids control is the suppression of algae growth since the hyacinth plant shades the water surface and prevents passage of sunlight to the water column.

Nitrogen removal. Plant uptake, ammonia volatilization, and nitrification/denitrification all contribute to nitrogen removal in hyacinth systems. Plant uptake, with plant harvest, can be an important removal pathway but nitrogen removal rates far in excess of plant uptake levels have been observed in a number of systems. A typical plant growth rate of about 220 kg/(ha · day) [196 lb/(acre · day)] (dry weight) would account for about 10 kg/(ha · day) [8.9 lb/(acre · day)] of nitrogen. The nitrogen removal actually observed at a number of systems was about 19 kg/(ha · day) [17 lb/(acre · day)] when the nitrogen loading ranged from 9 to 42 kg/(ha · day) [8 to 37 lb/(acre · day)].[30] The major factor responsible for this additional removal is believed to be nitrification/denitrification. The nitrifier organisms can flourish attached to the hyacinth roots, which provide oxygen, while adjacent microsites and the benthic layer will provide the anaerobic conditions and the carbon sources needed for denitrification. Nitrification/denitrification is more likely at a relatively shallow depth because the bulk of the wastewater has the opportunity for contact with the hyacinth root zone.

Pilot scale experiments with hyacinths and other aquatic plants in shallow containers [53 cm (21 in) deep] showed that overall nitrogen removal follows a first-order reaction rate.[21] The nitrogen removal observed was a function of plant density and temperature as shown by Eq. 5.1 and the rate constants in Table 5.4.[21]

$$\frac{N_e}{N_0} = \exp\,[-\,kt] \tag{5.1}$$

where N_e = total nitrogen in system effluent, mg/L
N_0 = total nitrogen in applied wastewater, mg/L
k = rate constant, dependent on temperature and plant density, days^{-1} (see Table 5.5 for values)
t = detention time in system, days

TABLE 5.4 Rate Constants for Eq. 5.1

Temperature and plant density	k, days^{-1}
Summer months	
Mean temperature 27°C ± 1°C	
Plant density, kg/ha (dry weight)	
3,920	0.218
10,230	0.491
20,240	0.590
Winter months	
Mean temperature 14°C ± 4°C	
Plant density, kg/ha (dry weight)	
4,190	0.033
6,690	0.023
20,210	0.184

Equation 5.1 is similar in form to the equation in Table 4.7, which estimates the nitrogen removal in pond systems. Either Table 4.7 or 4.8 can be used to estimate the nitrogen removal due to volatilization in hyacinth ponds. The results are not additive to Eq. 5.1 since the equation calculates the overall nitrogen removal which already includes a component for volatilization.

An analysis of data from the hyacinth systems listed in Table 5.3, as well as other sources, indicates that a correlation exists between nitrogen removal and the hydraulic loading on the basin surface. The relationship is described with Eq. 5.2, which is valid for a moderately dense (80 percent or more of basin surface covered with hyacinths) stand of plants with regular harvests to maintain optimum growth.

$$L_N = \frac{760}{(1 - N_e/N_0)^{1.72}} \tag{5.2}$$

where L_N = hydraulic loading, limited by nitrogen removal, $m^3/(ha \cdot day)$
N_e = nitrogen concentration required in system effluent, mg/L
N_0 = nitrogen concentration in influent to hyacinth basins, mg/L

In USCS units (L_N = million gals per day per acre) the equation becomes:

$$L_N = \frac{1}{(12.3)(1 - N_e/N_0)^{1.72}}$$

Phosphorus removal. The only significant removal pathway for phosphorus is plant uptake and that will usually not exceed 30 to 50 percent of the phosphorus present in typical municipal wastewaters. The removal will not even approach that range unless there is a careful vegetation management program involving frequent harvest. Maximum plant uptake of phosphorus may also require supplemental nitrogen fertilization since the ratio of nitrogen to phosphorus in typical wastewaters is significantly different from the balance required by the hyacinth plants (N:P = 6:1). As a result, there may be a nitrogen deficiency in the final basins of a hyacinth system and these plants cannot utilize the available phosphorus without additional nitrogen.

In typical systems where careful control and supplemental nutrients are not provided the phosphorus removal will probably not exceed 25 percent. Chemical precipitation with alum, ferric chloride, or other chemicals in a separate treatment step is recommended if high levels of phosphorus removal are a project requirement. Equation 5.3, derived from a number of operational systems can be used to estimate the potential for phosphorus removal in hyacinth basins. As with Eq. 5.2 it is valid when the basin surfaces are at least 80 percent covered with plants and there is a regular harvest.

$$L_p = (9353)\frac{P_e - 0.778P_0}{P_0 - P_e} \tag{5.3}$$

where L_P = hydraulic loading, limited by phosphorus removal, $m^3/(ha \cdot day)$
P_e = phosphorus concentration required in system effluent, mg/L
P_0 = phosphorus concentration in influent to hyacinth basins, mg/L
$[m^3/(ha \cdot day)]/(9353)$ = million gal/(day · acre)

Metals removal. Hyacinth systems are capable of high levels of metal removal. Although plant uptake can be significant the principal mechanisms are believed to be chemical precipitation and adsorption on substrate and on the plant surfaces. Mature plants will begin to slough root matter so any adsorbed material will then become part of the detritus or benthic sludge. In a study in Texas, Dinges[7] found that metals concentration in the bottom sediments exceeded the concentration in the living hyacinth plant tissue by at least an order of magnitude. This sediment consisted of a 2-year accumulation of biological solids as well as dead and sloughed plant material. The removal of

TABLE 5.5 Trace Element Removal by Water Hyacinths[14]

	Percent removal			
	With hyacinths		Without hyacinths	
Parameter	Batch	Continuous flow	Batch	Continuous flow
Arsenic	12	41	4	23
Boron	12	36	1	—
Cadmium	69	85	23	39
Mercury	70	92	60	93
Selenium	8	60	0	21

trace minerals observed in a 28-day batch experiment and in a 15-day continuous flow experiment are compared in Table 5.5.

Removal of trace organics. The removal of some organic priority pollutants has been measured in a pilot scale hyacinth basin system in San Diego, California. The hyacinth units in this case were used as a preliminary step ahead of ultrafiltration, reverse osmosis, carbon adsorption, and disinfection in a process intended to demonstrate the capability for complete water recycle and reuse. As shown on Table 5.6 excellent removal of trace organics was demonstrated in these hyacinth basins. The removal of trace organics is believed to be primarily due to decomposition of the compounds by bacterial action, although the plant itself can take up significant quantities of these materials.

Design considerations. Hyacinth systems can be designed for treatment of raw wastewater, primary effluent, upgrading of existing secondary treatment systems, or for advanced secondary or even tertiary treatment. As with other pond systems, the critical design parameter is the organic loading on the system.

If the project goal is secondary treatment the system design is essentially the same as given in Chap. 4 for a facultative pond. Table 5.7 presents a summary of the appropriate engineering criteria when hyacinths are used. The major function of the hyacinth plants in this case is the surface cover provided by the floating vegetation. This will

TABLE 5.6 Trace Organic Removal in Hyacinth Basins[31]

Parameter	Concentration, μg/L	
	Untreated wastewater	Hyacinth effluent*
Benzene	2.0	ND†
Toulene	6.3	ND
Ethylbenzene	3.3	ND
Chlorobenzene	1.1	ND
Chloroform	4.7	0.3
Chlorodibromomethane	5.7	ND
1,1,1 Trichloroethane	4.4	ND
Tetrachloroethylene	4.7	0.4
Phenol	6.2	1.2
Butylbenzyl phthalate	2.1	0.4
Diethyl phthalate	0.8	0.2
Isophorone	0.3	0.1
Naphthalane	0.7	0.1
1,4 Dichlorobenzene	1.1	ND

* Pilot scale system, 4.5 day detention time, 76 m^3/day flow, three sets of two basins each, in parallel, plant density 10–25 kg/m^2 (wet weight).

† ND = not detected.

TABLE 5.7 Suggested Criteria for Secondary Treatment with Hyacinth Ponds

Factor	Criterion
Effluent requirements	BOD <30 mg/L, SS <30 mg/L
Wastewater input	Untreated
Organic loading	
Entire system surface	50 kg/(ha · day) [45 lb/(acre · day)] BOD
First cell in system	100 kg/(ha · day) [90 lb/(acre · day)] BOD
Water depth	<1.5 m (5 ft)
Maximum area, single basin	0.4 ha (1 acre)
Total detention time	>40 days
Hydraulic loading	+200 m^3/(ha · day) [13,900 gal/(acre · day)]
Water temperature	>10°C (>50°F)
Basin shape	Rectangular, L:W > 3:1
Influent flow diffusers	Recommended
Mosquito control	Necessary
Harvest schedule	Seasonal or annual
Multiple cells	Essential, 2 sets of 3 basins, each recommended

prevent algal growth and contribute to BOD and SS removal. The performance of the hyacinth system will be significantly better than a comparable-sized facultative pond with an open water surface. In addition to new designs, hyacinth plants can be added to the final cells in existing facultative ponds to upgrade effluent quality to acceptable levels.

Multiple cells in pond systems are essential for proper hydraulic control (as described in Chap. 4), and are also important in hyacinth systems to ensure effluent quality during harvesting and other maintenance operations. A conservative approach to design, given in Table 5.7, divides the total treatment area required into two interconnected parallel rows of basins, with at least three basins in each set. This will allow temporary flow diversion for maintenance without disruption of overall performance.

Some states require the design of duplicate hyacinth systems, each capable of treating the design flow. It is therefore necessary to check with the appropriate regulatory authorities before proceeding with final project design.

Suggested engineering criteria for advanced secondary treatment using hyacinth ponds are given in Table 5.8. It is assumed in this case that at least primary treatment has been provided in a preliminary step. This could be achieved with a suitable aerobic or anaerobic pond, with conventional primary treatment, or with an Imhoff tank for small communities. It has been shown to be cost effective to provide supplemental aeration in these hyacinth systems to accelerate the treatment and allow increased loadings and shorter detention times. If aeration is not provided, the organic loadings should not exceed the values given

TABLE 5.8 Suggested Criteria for Advanced Secondary Treatment with Hyacinth Ponds

Factor	Criterion
Effluent requirements	BOD <10 mg/L, SS <10 mg/L, some nitrogen removal
Wastewater input	Equivalent to primary
Organic loading	
Entire system surface	100 kg/ha · day) [90 lb/(acre · day)] BOD
First cell surface	300 kg/(ha · day) [270 lb/(acre · day)] BOD
Detention time	>6 days
Aeration requirements	Design as partial mix aerated pond to meet O_2 needs (see Chap. 4), use submerged diffused aeration in first two cells of each set.
Water temperature	>20°C (68°F)
Water depth	<0.9 m (3 ft)
Hydraulic loading	<800 m³/(ha · day) [86,500 gal/(acre · day)]
Basin shape	Rectangular, L:W > 3:1
Influent flow diffuser	Essential
Effluent collection manifold	Essential
Single basin area	<0.4 ha (1 acre)
Mosquito control	Necessary
Harvest schedule	>monthly
Multiple cells	Essential, 2 interconnected parallel sets of 3 basins each

in Table 5.7. The shallower depth used in this case also allows the hyacinth plant to contribute more effectively to treatment than in the previous case. A tertiary hyacinth system, primarily for nutrient removal, can be an add-on to the system described in Table 5.7 or to any other secondary treatment process. Typical engineering criteria are described in Table 5.9. The use of the criteria in these tables is illustrated in the design examples that follow.

Example 5.1. Design a hyacinth system to produce secondary effluent with an untreated municipal wastewater as influent. Assume: design flow rate = 760 m^3/day; wastewater characteristics are BOD_5 = 240 mg/L, SS = 250 mg/L, TN = 25 mg/L, TP = 15 mg/L; and critical winter temperature >20°C (68°F). Effluent requirements: BOD_5 = <30 mg/L, SS <30 mg/L.

solution

1. Determine BOD loading:

$$(240\ \text{mg/L})(760\ \text{m}^3/\text{day})(10^3\ \text{L/m}^3)(1\ \text{kg}/10^6\ \text{mg}) = 182.4\ \text{kg/day}$$

2. Determine basin surface areas based on criteria in Table 5.7: 50 kg/(ha · day) BOD for entire area, 100 kg/(ha · day) BOD for first cell.

$$\text{Total area required} = \frac{182.4\ \text{kg/day}}{50\ \text{kg/(ha} \cdot \text{day)}} = 3.65\ \text{ha}$$

$$\text{Surface area of first cells} = \frac{182.4\ \text{kg/day}}{100\ \text{kg/(ha} \cdot \text{day)}} = 1.82\ \text{ha}$$

3. Use two primary cells, each 0.91 ha in area; with $L:W = 3:1$, the dimensions at the water surface will be:

$$A = L/W = (L)(L/3) = (L^2/3) = (90.91 \text{ ha})(10{,}000 \text{ m}^2/\text{ha})$$

$$L = 165 \text{ m and } W = 165/3 = 55 \text{ m}$$

4. Divide the remaining required area into two sets of two basins each to produce a total system with two parallel sets with three basins each.

$$\text{Total area final cells} = 3.65 \text{ ha} - 1.82 \text{ ha} = 1.83 \text{ ha}$$

$$\text{Individual cells} = 1.83 \text{ ha}/4 = 0.46 \text{ ha}$$

$$L^2 = (3)(0.46 \text{ ha})(10{,}000 \text{ m}^2/\text{ha})$$

$$L = 117 \text{ m and } W = 117 \text{ m}/3 = 39 \text{ m}$$

5. Allow 0.5 m for sludge storage and assume a 1 m "effective" water depth for treatment; total pond depth = 1.5 m. Use 3:1 side slopes, and use Eq. 4.15 to determine the treatment volume.

$$V = [(L \times W) + (L - 2sd)(W - 2sd) + 4(L - sd)(W - sd)]d/6$$

Primary cells:

$$V = [(165)(55) + (165 - 2 \times 3 \times 1)(55 - 2 \times 3 \times 1) + 4(166 - 2 \times 1)(55 - 2 \times 1)]1/6$$

$$V = 8570 \text{ m}^3$$

Final cells:

$$V = [(117)(39) + (117 - 2 \times 3 \times 1)(39 - 2 \times 3 \times 1) + 4(117 - 2 \times 1)(39 - 2 \times 1)]1/6$$

$$V = 4208 \text{ m}^3$$

TABLE 5.9 Suggested Criteria for Tertiary Treatment with Hyacinth Ponds

Factor	Criterion
Effluent requirements	BOD <10 mg/L, SS <10 mg/L, TN and TP <5 mg/L
Wastewater input	Secondary effluent
Organic loading	
Surface of entire system	<50 kg/(ha · day) [45 lb/(acre · day)] BOD
Surface of first cell	<150 kg/(ha · day) [135 lb/(acre · day)] BOD
Water depth	≪0.9 m (≪3 ft)
Maximum area, single cell	<0.4 ha (<1 acre)
Detention time	6 days or less, depending on depth
Hydraulic loading	<800 m³/(ha · day) [86,500 gal/(acre · day)]
Basin shape	Rectangular, L:W > 3:1
Water temperature	>20°C (>68°F)
Mosquito control	Necessary
Influent flow diffuser	Essential
Effluent collection manifold	Essential
Harvest schedule	Mature plants, every few weeks
Multiple basins	Essential, same as Table 5.8

6. Determine the hydraulic detention time in the "effective" treatment zone:

$$\text{Primary cells: } t = \frac{8570 \text{ m}^3}{(760 \text{ m}^3/\text{day})/2} = 22.5 \text{ days}$$

$$\text{Final cells; } t = \frac{(2)(4208 \text{ m}^3)}{(760 \text{ m}^3/\text{day})/2} = 22.1 \text{ days each}$$

$$\text{Total detention time} = 22.5 + 22.1 + 22.1 = \text{days}$$

$$> 50 \text{ days,} \quad \text{OK}$$

7. Check hydraulic loading:

$$\frac{760 \text{ m}^3/\text{day}}{3.65 \text{ ha}} = 208 \text{ m}^3/(\text{ha} \cdot \text{day}) \quad \text{OK}$$

8. Estimate nitrogen removal with Eq. 5.2 to be sure sufficient nitrogen is present to sustain growth in the final cells and to determine harvest frequency. Rearrange Eq. 5.2:

$$\left(1 - \frac{N_e}{N_0}\right)^{1.72} = \frac{760}{L_N}$$

$$N_e = N_0\left[1 - \left(\frac{760}{L_N}\right)^{1/1.72}\right]$$

$$= 25\left[1 - \left(\frac{760}{208}\right)^{1/1.72}\right]$$

This would predict a negative nitrogen concentration in the effluent which is not possible. The basic equation would predict essentially complete removal at a hydraulic loading of 760 m^3/(ha · day). Since the loading for this example is only 208 m^3/(ha · day) it is reasonable to expect 5 mg/L of nitrogen in the final effluent or less. Since the nitrogen will not be at optimum growth levels in this system an annual harvest is suggested. An influent flow diffuser in each of the primary cells is recommended to properly distribute the untreated influent.

Example 5.2. Design a hyacinth system to produce advanced secondary effluent on a site with limited available area. Design flow = 760 m^3/day; wastewater characteristics are BOD_5 = 240 mg/L, SS = 250 mg/L, TN = 25 mg/L, TP = 15 mg/L; and winter water temperatures >20°C (68°F). Effluent requirements: BOD_5 = <10 mg/L, SS = <10 mg/L, TN <10 mg/L. Assume 80 percent plant coverage is maintained on the basins and routine monthly harvests are included.

solution

1. Since the site area is limited, space is not available for preliminary treatment in a pond unit. Use Imhoff tanks for primary treatment and supplemental diffused aeration in the hyacinth ponds to minimize area requirements. The Imhoff tank has the added advantage for this relatively small flow in that separate sludge digestion is not required.
2. Design the Imhoff tank.
   Typical criteria:

$$\text{Sedimentation detention time} = 2 \text{ hr}$$

$$\text{Surface loading} = 24 \text{ m}^3/(\text{m}^2 \cdot \text{day})$$

$$\text{Over flow weir loading} = 600 \text{ m}^3/(\text{lin m/day})$$

Surface area for scum = 20% of total surface

Sludge digestion volume = 0.1 m^3/capita for the population served, or about 33% of total tank volume.

$$\text{Minimum sedimentation area needed} = \frac{760\ m^3/day}{24\ m^3/(m^2 \cdot day)} = 31.7\ m^2$$

$$\begin{aligned}\text{Minimum total surface area} &= \text{sedimentation} + \text{scum}\\ &= (1.20)(31.7\ m^2)\\ &= 38\ m^2\end{aligned}$$

A typical tank might be 8 m long and 5 m wide. In this case the central sedimentation chamber might be 4 m wide with open channels on each side, about 0.5 m wide, for scum accumulation and gas venting. The slotted, sloping bottom (bottom walls sloped at 5:4) would have to be about 3 m deep to provide the necessary 2-h detention time. The total depth of the hopper bottomed tank might be 6 to 7 m including an allowance for freeboard and the sludge digestion volume.

A properly maintained Imhoff tank can achieve about 47 percent BOD removal and up to 60 percent SS removal.[2] Assuming no nitrogen or phosphorus losses the primary effluent for this example would be:

$$BOD_5 = (240\ mg/L)(0.53) = 127\ mg/L$$

$$SS = (250\ mg/L)(0.40) = 100\ mg/L$$

$$TN = 25\ mg/L$$

$$TP = 15\ mg/L$$

3. The BOD loading on the hyacinth basins would be:

$$(127\ mg/L)(760\ m^3/day)(10^3\ L/m^3)(1\ kg/10^6\ mg) = 96.5\ kg/day$$

4. Determine the basin surface areas. From Table 5.8 the allowable organic loading on the entire area would be 100 kg/(ha · day) and up to 300 kg/(ha · day) on the first cell.

$$\text{Total surface area required} = \frac{96.5\ kg/day}{100\ kg/(ha \cdot day)} = 0.97\ ha$$

$$\text{Surface area for primary cells} = \frac{96.5\ kg/day}{300\ kg/(ha \cdot day)} = 0.32\ ha$$

5. Use two primary cells in parallel, each 0.16 ha in area, use rectangular shape with $L:W = 3:1$; so: $L = 69$ m, $W = 23$ m.

6. Divide the remaining area into two sets of two cells each to produce two parallel sets with a total of three basins each.

$$\frac{0.97\ ha - 0.32\ ha}{4} = 0.16\ ha\ \text{each}$$

With $L:W = 3:1$; $L = 69$ m and $W = 23$ m

7. Allow 0.5 m for sludge storage and assume 0.6 m for "effective" water depth for treatment in basins with 3:1 side slopes. Determine treatment volumes (see equation in Example 5.1).

All basins are the same size, $V = 855\ m^3$

$$\text{Detention time single basin} = \frac{855\ m^3}{(760\ m^3/day)/2} = 2.25\ days$$

Total detention time = (2.25)(3) = 6.75 days >6 days OK

8. Check hydraulic loading = 760 m^3/day/0.97 ha = 783 m^3/(ha · day) OK
9. Determine nitrogen removal (see Example 5.1 for basic equation)

$$N_e = N_0\left[1 - \left(\frac{760}{L}\right)^{1/1.72}\right]$$
$$= 25[1 - (0.97)^{1/1.72}]$$
$$= 0.5 \text{ mg/L} \quad \text{OK}$$

10. Design a partial mix diffused aeration system for the first two hyacinth basins in each set. Assume that the required oxygen is double the organic loading, the air contains about 0.28 kg/m^3 oxygen, and the aeration efficiency in the shallow basins is about 8 percent (usually 16 percent or more at normal lagoon depths).

$$\text{Total air required} = \frac{(2)(\text{BOD, mg/L})(Q\text{, L/day})(10^{-6}\text{ mg/kg})}{(E)(0.28\text{ kg/m}^3)(86{,}400\text{ s/day})}$$
$$= \frac{(2)(127\text{ mg/L})(760 \times 10^3\text{ L/day})(10^{-6})}{(0.08)(0.28)(86{,}400)}$$
$$= 0.1\text{ m}^3\text{/s}$$

Manufacturers' literature should be used to select the specific aeration devices. In this case, about $\frac{2}{3}$ of the aeration capacity is split between each of the primary hyacinth cells and the remaining $\frac{1}{3}$ is divided equally for the second cells in each set. Typical submerged aeration tubing can supply about 2.5 × 10^{-3} m^3/min of air per meter of tubing [0.027 ft^3/(min/lin ft)]. Determine length and location of aeration tubing:

$$\text{Total length} = \frac{(0.1\text{ m}^3\text{/s})(60\text{ s/min})}{(2.5 \times 10^{-3}\text{ m}^3\text{/min})} = 2400\text{ m}$$

$$\text{In primary basins} = \frac{(2400)(0.667)}{2} = 800\text{ m each}$$

$$\text{Number of aeration lines} = \frac{\text{tubing length}}{\text{basin width}} = \frac{800\text{ m}}{23\text{ m}} = 35\text{ each}$$

Space these aeration lines on 2 m centers in the primary basins.

$$\text{In second basins} = \frac{(2400)(0.333)}{2} = 400\text{ m each}$$

$$\text{Number of lines} = \frac{400}{23} = 17\text{ in each basin}$$

Space these at 4 m center to center for the full length of the basin.

11. An inlet diffuser system or sprinklers are essential for the primary cells to ensure uniform distribution of influent. The use of *Gambusia* fish or other biological or chemical agents are necessary for mosquito control. Harvest of plants should be conducted about every 3 to 4 weeks with not more than 20 percent of the plant cover removed at any one time.

12. The treatment system designed in this example will provide better performance than the system developed in Example 5.1 on less than $\frac{1}{3}$ of the land area. The major reasons are the use of the Imhoff tank for primary treatment and aeration in the first two basins in each set. In locations where land is limited or very expensive this approach to treatment might still be cost effective even when just secondary level treatment is required.

Structural elements. The small individual basins suggested in Tables 5.7 to 5.9 are recommended to facilitate harvesting of the hyacinth plants at small to moderate sized systems. The long narrow configuration suggested is for hydraulic control and ease of harvesting. The width of the basin will depend on the capabilities of the harvesting equipment. If the system is drained on an annual basis for plant removal, an access ramp is needed in each basin; the basin width being not especially critical since a front end loader or similar equipment can be used for basin cleaning. The higher rate systems with more frequent harvests require access for floating devices or they must have roads on the dikes for equipment access. A typical drag-line bucket might have a 9-m (30-ft) range so a basin might be 15 to 18 m (50 to 60 ft) wide and designed for harvesting from both sides.

The use of multiple influent points is recommended for secondary systems and is essential for the higher rate, high performance systems. This is to ensure proper wastewater distribution and effective use of the entire treatment volume and to maintain aerobic conditions throughout the basin. Sprinklers can also be used for influent distribution and have the added advantage of providing some frost protection during cold weather periods. Experience with long rectangular hyacinth cells with a single inlet has demonstrated that most of the solids and BOD removal occur near the headworks. This can create undesirable anaerobic conditions in this area which can result in odor problems and ineffective mosquito control as well as being an ineffective use of the total treatment volume. The influent works should be designed to uniformly apply wastewater over the initial $\frac{1}{3}$ to $\frac{1}{2}$ of the surface area in the first hyacinth covered basin in each set, when the basins are square or rectangular. Other inlet/outlet configurations, suggested by Tchobanoglous[28] to ensure better utilization of the basin area, are shown in Fig. 5.2.

An effluent manifold spanning the entire basin width in the typical rectangular cells is also recommended to avoid "dead" spots and the resulting ineffective treatment near the outlet. These manifolds are suggested for interbasin transfer and for the final effluent discharge. An alternative, as shown in Fig. 5.3 is to narrow the channel width as the outlet point is approached. This will serve to increase the flow velocity toward the outlet and thereby eliminate the "dead" spots. These manifolds or single discharge points should be at the water surface in all of the basins to ensure that all water is brought up into contact with the hyacinth roots prior to discharge. In relatively wide basins changing the width near the outlet will not be effective. The approach in this case, as shown in Fig. 5.3, is to slope the basin bottom upward in the discharge zone to create a shallow depth to ensure contact with the plants. Screening or a baffle is necessary ahead of the

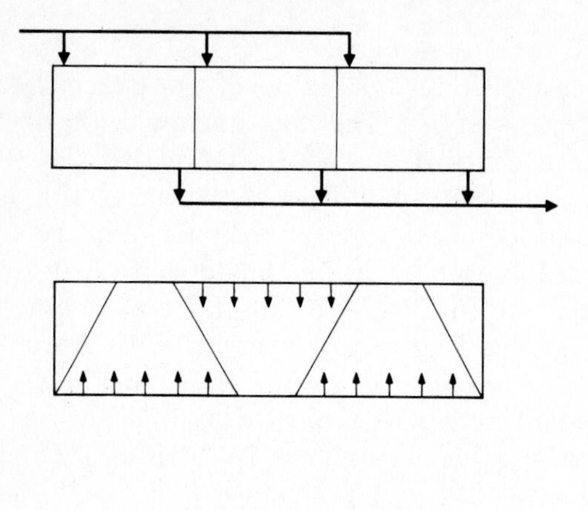

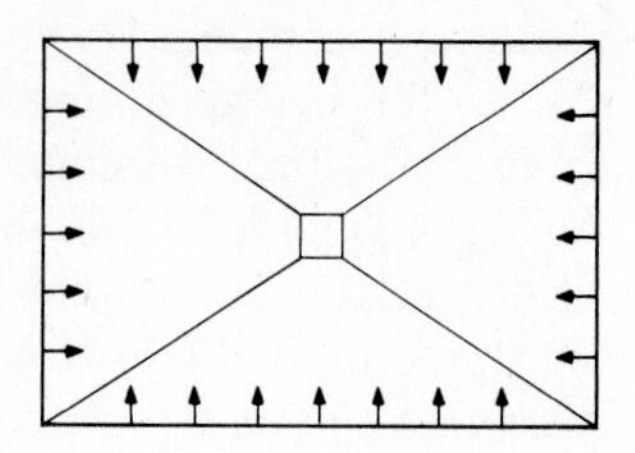

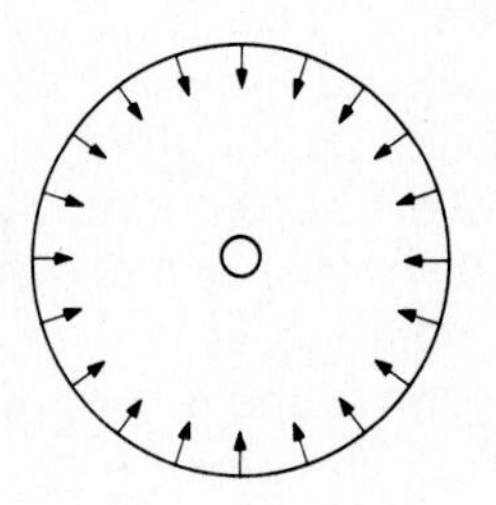

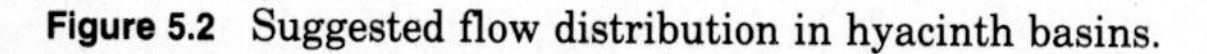

Figure 5.2 Suggested flow distribution in hyacinth basins.

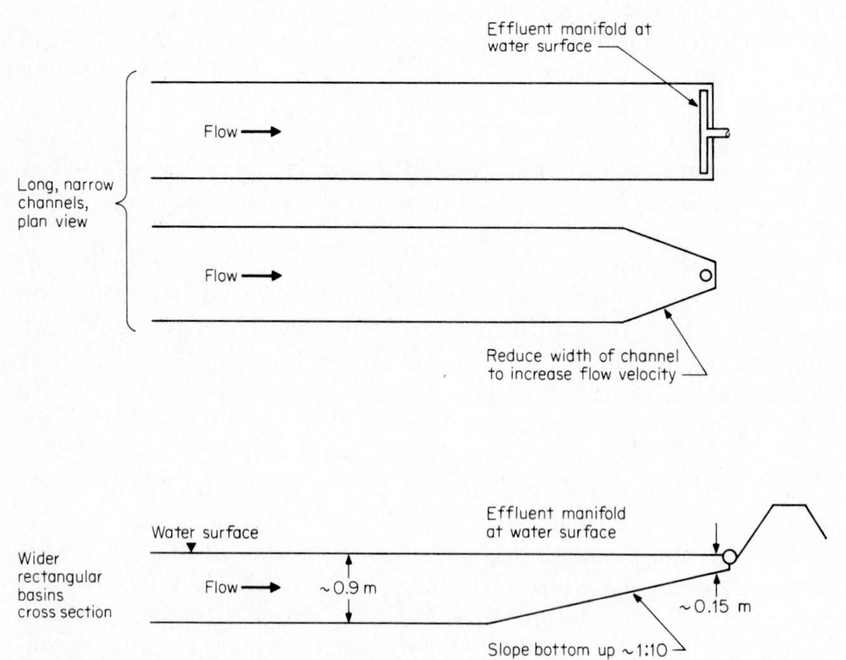

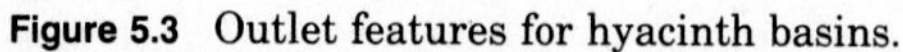

Figure 5.3 Outlet features for hyacinth basins.

manifold or outlet to prevent loss of hyacinth plants with the effluent.

Long narrow channels can be constructed with concrete or other structural sidewalls and a lined bottom. The construction of wider basins is essentially the same as the pond systems described in Chap. 4. Exterior dikes should be about 3 m (10 ft) wide at the top to permit vehicle movement; side slopes should be 3:1 and the dike constructed to provide about 0.5 m (1.6 ft) of freeboard above the design water surface.

State or local regulations will control the degree of permeability allowed in the basin bottom. It is likely that lining or some other impermeable barrier will be required in most cases if permeable soils are dominant on the site (see Sec. 4.12 for further discussion). The bottom of the basin should be smooth and constructed at a slight grade (0.5 percent) toward the outlet to facilitate drainage. Construction of a sump in the outlet area is also suggested for the same purpose.

The optimum water depth in a hyacinth basin depends on the intended function of the vegetation and on the desired effluent quality. The depth is not critical if the major purpose of the hyacinth plant is surface shading to prevent algae growth. A relatively shallow depth is desirable when the plants are expected to provide significant nutrient removal. The optimum water depth in a carefully managed high-rate system might range from 0.3 m (1 ft) in the first basin to 0.45 m (1.5 ft) in the final basins. A greater depth is used in the final cell since the hyacinth roots will be longer when fewer nutrients are present in the water. The discharge zone in these final basins might then be reduced to the 0.15 m (0.5 ft) shown in Fig. 5.3 to ensure full contact with the plants prior to final discharge. A design using these shallow depths (0.3 to 0.5 m) should be able to reduce the maximum design detention times given in Tables 5.8 and 5.9. A pilot scale test is suggested for large scale projects to optimize these design parameters.

Operation and maintenance. The major operational concerns are control of mosquitos and odors, vegetation management, sludge removal, plant harvest, and the disposal or utilization of the harvested materials and sludge. Other requirements include all of the routine activities common to the operation and maintenance of pond systems, which are the same for hyacinth systems and the lagoons described in Chap. 4.

A special concern with hyacinth ponds is the potential for significantly higher evapotranspiration (ET) from the plants as compared to evaporation from an open water surface under the same climatic conditions. Several research efforts have shown the ET rates for hyacinths are about 3 times the evaporation rate for open water.[22] The ET losses calculated for hyacinth basins at Kissimmee, Florida were about 20 to

33 L/(m^2 · day) [0.5 to 0.8 gal/(day · ft^2), which was about 3 times the expected pan evaporation rate for the area.[1] This can be a critical factor when hyacinth systems are planned for arid climates where excessive water losses are not desirable.

Mosquito control. Control with chemical sprays is not practical because the mosquito larvae in hyacinth ponds are at the water surface beneath the leaf canopy. Several pilot systems in California were closed because of mosquito problems. An effective control method is to stock each basin with *Gambusia* or similar small surface feeding fish that prey on the mosquito larvae. These fish will not tolerate anaerobic conditions and will not enter water zones with low oxygen levels. Avoiding such anoxic conditions near the basin inlets is one of the reasons for installing influent diffusers. These small tropical fish will not tolerate low water temperatures either. If a seasonal hyacinth operation is planned it will be necessary to restock the basins with both plants and fish at the start of the warm weather period. A typical initial stocking rate for the *Gambusia* fish is about 7000 to 12,500/ha (2800 to 5000/acre) of surface area. Other species used for mosquito control include goldfish (*Carassius auratus*), frogs (*Hyla* sp.), and grass shrimp (*Palemonetes kadiakensis*). If algae control is necessary, Blue Tilapia (*Tilapia aureaus*), Sailfin Mollies (*Poccilia latipinna*), and Japanese Koi (*Cyprinus* sp.) can be used. The hyacinth basins in the system constructed at Austin, Texas incorporate small fenced-off zones to maintain an open water surface and sufficient aeration from natural sources to support the *Gambusia* fish.[8] The basins should be stocked with fish a few weeks prior to stocking with the hyacinth plants.

Odor control. Since the floating mat of plants suppresses algae and prevents wind-induced surface reaeration the only source of oxygen is from the phytosynthetic respiration of the hyacinth plants. In unaerated basins this natural source of oxygen will not be enough to sustain general aerobic conditions with moderate to high BOD loadings. If the wastewater contains more than 30 mg/L sulfates, the anaerobic conditions will probably result in objectionable hydrogen sulfide odors. This is another reason for the broad distribution of the influent in at least the first basin in a hyacinth system. Supplemental aeration for odor control may still be necessary in these primary basins at night and during other phytosynthetically inactive periods.

Vegetation management. The degree of vegetation management required depends on the water quality goals of the project and a choice between harvesting plants or frequent sludge removal. A frequent plant harvest may be necessary to sustain a significant level of phosphorus removal but is not necessary for nitrogen removal. Studies in Florida have shown nitrogen removal rates to be 2 to 3 times higher in unharvested basins as compared to frequently harvested ones.

When the plant density on the water surface exceeds about 25 kg/m^2 (5 lb/ft^2) (wet weight) sloughing of root material commences. This accumulation of plant detritus on the basin bottom will, after a few months, exceed the mass of settled wastewater solids. One approach, recommended by the State of Texas, uses an annual draining and cleaning of each basin instead of regular plant harvest. All of the plants as well as the benthic sludge are removed and the basin is then refilled and restocked with new plants. Systems in Florida and elsewhere have adopted a more frequent plant harvest and a less frequent basin cleaning.

Frequent harvests are considered necessary to keep the plants at the optimum growth stage to ensure optimum phosphorus removal. In these cases, the plant density is maintained between 10 and 25 kg/m^2 (2 and 5 lb/ft^2) (wet weight). One technique for monitoring plant density is to use mesh bottomed floating baskets about 1 m on a side. The basket is periodically lifted out of the basin and weighed to determine the wet weight density of the plant cover. System designs based on Wolverton's research[31–33] recommend wet weight plant densities from 12 to 22 kg/m^2 (2.5 to 4.6 lb/ft^2) for optimum treatment with loosely packed plants with 80 to 100 percent surface coverage. An initial plant stocking rate of 1.8 kg/m^2 (0.37 lb/ft^2) has been used in Florida.[1]

Nutrient and micronutrient deficiencies have also been observed in the final basins of hyacinth systems. Plant chlorosis (leaf yellowing) due to iron deficiency has occurred in several systems in Florida. The problem was corrected with the addition of ferrous sulfate at a rate sufficient to maintain the iron concentration in the water at about 0.3 mg/L.

Insect infestations can cause major damage to the plants. The caterpillar stage of the moth *Sameodes allijuttales* and the weevils *Neochetina eichornia* and *N. bruchi* attack the plant stolon and the leaves, respectively. The weevils seem to be more active when the plants are under density stress and the moths are more likely to be a problem with hot, dry weather conditions. The life cycle for the weevils is about 60 days, with peaks in the spring and fall. Spot harvests may be an effective control in the early stages, and the insecticide Sevin has been used for major infestations.[15]

The hyacinth plant does not tolerate cold, and even short periods of freezing weather can destroy this important component in the treatment process. The 1.6-ha (4-acre) hyacinth system in Austin, Texas is entirely covered with a greenhouse structure to permit year-round operation. Other plant types are also being investigated for combined use with hyacinths. One possibility is the pennywort (*Hydrocotyle umbellata*), which is more cold tolerant than the hyacinth and also has a higher oxygen transfer rate to the root zone. Combined hyacinth-pennywort systems in Florida perform better and more reliably than monoculture units with either of the plants.[5]

Sludge removal. The benthic sludge consisting of wastewater solids and plant detritus must eventually be removed from all hyacinth systems. An annual cleaning of the primary cells in very shallow high-rate systems may be needed even with frequent harvests. The secondary and tertiary cells in these systems may only need cleaning every 2 to 3 years. The deeper hyacinth systems with regular harvest, which are designed for secondary treatment only, should be cleaned on a 5-year cycle. Systems with no harvest, or those operated on a seasonal basis should be cleaned on an annual basis. The cleaning method will depend on the basin configuration and its construction materials. Large basins constructed of compacted earth, concrete, asphalt, or protected membrane liners could use conventional front-end loaders for sludge removal from the drained basins. Small basins could use float-supported suction pumps or dredges. Since the sludge will contain wastewater solids, its subsequent treatment and disposal must comply with local regulatory practices.

Harvest procedures. The harvest frequency may range from a few weeks to a month or more depending on the level of nutrient removal required. If a complete harvest is needed for insect control, frost damage, or other reasons, restocking at a density of about 7 kg/m^2 (1.5 lb/ft^2) (wet weight) will promote optimum growth and rapid coverage of the basin.[15]

A number of methods have been tried for harvest of the hyacinth plants including front-end loaders, draglines or backhoes equipped with clamshell buckets or weed buckets, conveyors, conveyor-chopper systems, chopper pumps, rakes, and boats. The equipment selected should be able to easily reach any part of the hyacinth basins to allow selective harvests of mature plants. Wolverton[33] compared conveyor-choppers, a conveyor with a pusher boat, and a dragline equipped with a modified clamshell bucket. The conveyor-pusher boat and the dragline had about the same production rates: 418 m^2/h (4500 ft^2/h) with a plant density of about 22 kg (wet wt)/m^2. The dragline was recommended for its greater mobility and reliability. Modified truck or tractor mounted backhoe devices have also been successfully used. Instead of the normal bucket attachment basketlike tines are placed at the end of the articulating arm. These devices are suitable for small- to moderate-sized systems with channel-type designs. The limiting factor for the economics of the operation is the cost of transport from the basin to the disposal/utilization site. A typical 12 m^3 (16 yd^3) dump truck can hold about 5 to 7 metric tons (6 to 8 tons) of wet hyacinth plants.

Larger scale systems designed for both wastwater treatment and for biogas production will require in-basin harvesting techniques and a more efficient transport system than the trucks used at the smaller operations. Recent developments in Florida utilize winch-operated floats or a floating pusher vehicle to move the plants to the onshore chopper and progressive cavity pump, which can then deliver the chopped plants as a slurry (about 4 percent) solids directly to the biogas digestor. This equipment can harvest 9 metric tons (10 tons) of plants per hour at an approximate cost of $2.00/metric ton ($2.30/ton).

The hyacinth system in Coral Springs, Florida[27] reports the best effluent water quality performance with "loosely packed" hyacinths on the water surface. A four-week harvest schedule is used and not more than 15 to 20 percent of the plants are taken at one time. A truck-mounted dragline with a weed bucket is used, with a dump truck for transport of the harvested material. A production rate of 700 m^2 (7300 ft^2) per hour was reported with this equipment. The harvest was reported in volumetric units and was 2.7 m^3 per 100 m^2 of basin surface (3.5 yd^3/1000 ft^2). If a plant surface density of 22 kg/m^2 (4.6 lb/ft^2) is assumed, the wet unit weight of the harvested plants would have been about 815 kg/m^3 (51 lb/ft^3).

Hyacinth disposal or utilization. Since the hyacinth plants are about 95 percent water, an intermediate drying step is usually employed prior to disposal or utilization of the harvested material at the smaller systems. Preliminary grinding, chopping, and pressing have been tried to accelerate the drying process. Covered solar drying racks have also been used, but the most common approach is to use a small open area adjacent to the basins for spreading and air drying of the whole harvested plants to the desired moisture content. The solar drying racks used in Florida have a 5-day drying cycle to reach a moisture content of 20 percent,[26] while an open bed might require 2 to 3 weeks to reach the same level in the same climate.

The dried plants can be disposed of in a landfill, or elsewhere, as permitted by the local regulatory authorities. If the wastewater has very high metal concentrations it may be advisable to check the metal content of the dried plant to ensure that the levels do not exceed permit allowances for disposal/utilization (see Chap. 8 for further discussion of these limits).

The simplest approach for beneficial reuse of the harvested materials is to compost the semidry hyacinths and then use that material as a soil conditioner/fertilizer. Anaerobic digestion of the plants and sludge for methane production and processing of the plants for animal feed have been shown to be technically feasible but marginally cost effective. Recent demonstrations in Florida using a 2:1 mixture of hyacinths and sludge in a vertical flow nonmixed anaerobic reactor have produced high quality methane in a cost-effective process.[17] The major factor is the novel reactor design which does not require the mixing energy used in conventional anaerobic digestors. There may not be sufficient plant production to sustain routine operation of these more complex processes at wastewater flows less than 3800 m^3/day (1 million gallons/day). The composting option is the best-suited option for smaller systems.

Duckweed

Duckweed, in the genera *Lemna* sp., *Spirodela* sp., and *Wolffia* sp., have all been tested for pollutant removal or used in wastewater treatment systems. These are all small, green freshwater plants with a leaflike frond a few millimeters in width with a short root usually less than a centimeter in length. The morphology of the plant is shown on Fig. 5.4.

These duckweeds are the smallest and the simplest of the flowering plants and have one of the fastest reproduction rates. A small cell in the frond divides and produces a new frond; each frond is capable of producing at least 10 to 20 more during its life cycle.[12] *Lemna* sp. grown in wastewater effluent (at 27°C) doubles in frond numbers, and therefore the area covered, every 4 days. It is believed that duckweed can grow at least twice as fast as other vascular plants. The plant is essentially all metabolically active cells with very little structural fiber.

Duckweed, like hyacinths, contains about 95 percent water; the composition of the plant tissue is given in Table 5.10. A comparison of the values in Tables 5.10 and 5.2 indicate that duckweed contains at least twice as much protein, fat, nitrogen, and phosphorus as hyacinth. Sev-

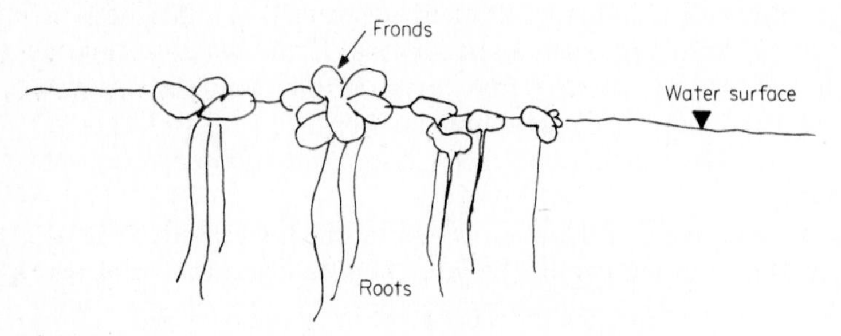

Figure 5.4 Morphology of duckweed plants.

TABLE 5.10 Composition of Duckweeds Grown in Wastewater[13]

Constituent	Percent of dry weight	
	Range	Average
Crude protein	32.7–44.7	38.7
Fat	3.0–6.7	4.9
Fiber	7.3–13.5	9.4
Ash	12.0–20.3	15.0
Carbohydrate	—	35.0
Kjeldahl nitrogen (as N)	4.59–7.15	5.91
Phosphorus (as P)	0.80–1.8	1.37

eral nutritional studies have confirmed the value of duckweed as a food source for a variety of birds and animals.[12]

Duckweeds are more cold tolerant than hyacinths and are found throughout the world. A minimum temperature of 7°C (45°F) has been suggested as the practical limit for growth of duckweeds.[16] As shown in Fig. 5.5 the range for a year-round duckweed treatment system is slightly greater than shown in Fig. 2.1 for hyacinths, but seasonal duckweed systems operating 6 months per year should be possible for most of the United States.

Performance expectations. Duckweed systems are capable of high levels of BOD and SS removal and significant levels of metal and nutrient removal. As compared to hyacinths the duckweed plant plays a less direct role in treatment due to its small size. The lack of an extensive root zone provides very little substrate for attached microbial growth. A number of studies have used wastewater to produce duckweed for animal feed[7] but there is very limited experience to date using the plants as a component in a system designed primarily for wastewater treatment.

The growing plants will form a single layer completely covering the water surface, then some species will grow on top of others. Their small size makes the plants susceptible to the wind; initially this may result in part of the basin being uncovered but the long-term effect is a thick mat of plants covering the entire basin. This mat is still susceptible to the wind, so floating booms or cells are usually used to hold the plants in place. The formation of this mat is probably the most significant contribution of the duckweed plant to wastewater treatment. This surface cover prevents algae growth, stabilizes pH, and enhances sedimentation but is also likely to result in anaerobic conditions due to the relatively low phytosynthetic oxygen production from the small plants. The plant can flourish under anoxic conditions but the rate of

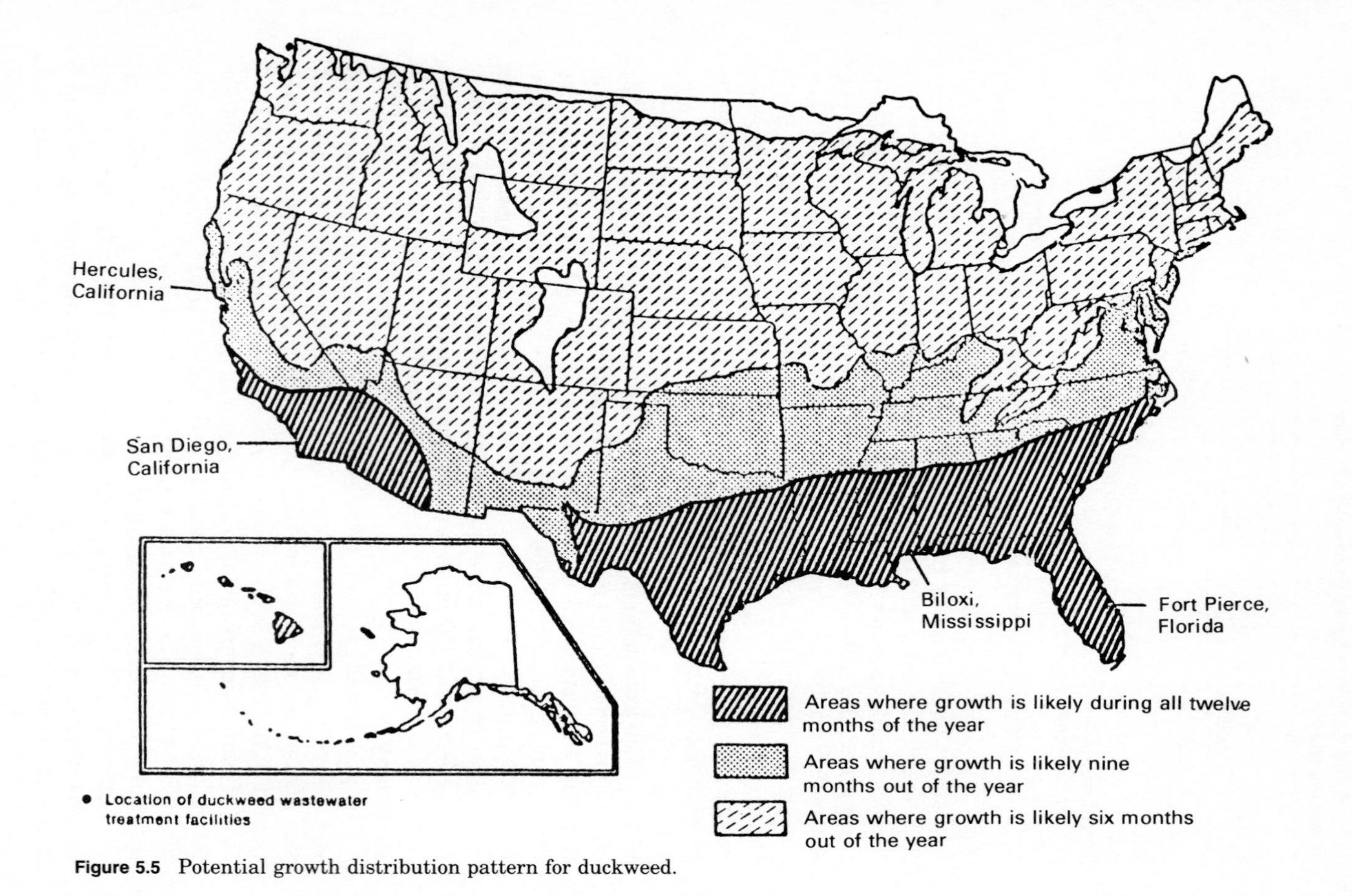

Figure 5.5 Potential growth distribution pattern for duckweed.

biological activity in the water will proceed at lesser rates than in an aerobic environment.

BOD removal. The major factors responsible for BOD removal in a duckweed system are the same as described in Chap. 4 for facultative stabilization ponds. The duckweed plants create the environment for treatment but contribute very little directly to the removal of BOD. Wolverton[33] has reported on the performance of a duckweed-covered basin (following an aerated cell) near Biloxi, Mississippi. The organic loading on this 22-day detention time basin was about 24 kg/(ha · day) [21 lb/(acre · day)], which is near the low end of the range for conventional facultative ponds. The final effluent from this basin contained about 15 mg/L of BOD and was anaerobic.

Suspended solids removal. The removal of suspended solids in duckweed basins is due to the same factors described above for BOD removal. Suspended solids removal in a duckweed-covered basin should be more effective as compared to a conventional stabilization pond due to the lack of algae and the improved quiescent conditions under the surface mat. The final SS concentration from the Cedar Grove system discussed in the previous section averaged 14 mg/L.

Nitrogen removal. Plant uptake, ammonia volatilization, and nitrification/denitrification are believed to contribute to nitrogen removal in a duckweed basin. Laboratory scale experiments indicate the plants contribute significantly but only account for about 25 percent of the nitrogen removed in the system.[21] Other experiments at the same scale [water depth 41 cm (16 in)] indicate that overall nitrogen removal follows a first-order reaction rate.[20] The nitrogen removal was influenced by the plant species, the plant density, and the temperature. Equation 5.1, with the rate constants in Table 5.11, can be used to estimate overall nitrogen removal in a duckweed system. Frequent harvests are necessary to sustain high levels of nitrogen removal. Since the duckweed plant has essentially no root zone the nitrification/denitrification reactions described previously for hyacinths cannot occur in these systems.

Phosphorus removal. Data from small-scale laboratory studies in Florida indicate that plant uptake by duckweed was responsible for about 30 percent of the phosphorus removed in the summer and only 10 percent during the winter.[21] If significant phosphorus removal is a project requirement the use of chemical precipitation with alum, ferric chloride, or other chemicals in a separate treatment step is suggested.

TABLE 5.11 Rate Constants for Estimating Nitrogen Removal with Eq. 5.1 in a Duckweed System[20]

Temperature and plant density	k, days^{-1}
Summer months	
Mean temperature 27°C ± 1°C	
Plant density, kg/ha (dry wt)	
Lemna minor	
73	0.074
131	0.011
Winter months	
Mean temperature 14°C ± 4°C	
Plant density, kg/ha (dry wt)	
Lemna minor	
40	0.028
67	0.012

Metals removal. Plant uptake of metals plays a lesser role in duckweed systems than described previously for hyacinths. The major removal mechanisms are chemical precipitation and ultimately incorporation in the benthic sludges.

Design considerations. The use of duckweeds for wastewater treatment is not as well developed as hyacinth systems and experience to date has been limited to laboratory and pilot scale research and field demonstrations. This experience suggests that the major function of the duckweed plants on lagoons of conventional depth is to provide a surface cover on the pond, rather than contributing directly to removal of pollutants.

The design of a duckweed pond system should, at this stage of development, follow the conventional design procedures for facultative ponds as presented in Chap. 4. Effluent from a duckweed covered system should exceed performance expectations for BOD, SS, and nitrogen removal as compared to a conventional pond system with an open water surface. The effluent from such a system is likely to be anaerobic and post aeration of some type may be necessary. The duckweed pond at Cedar Lake uses turbulent flow during a 0.9-m (3-ft) drop to aerate the final effluent.[31]

Wolverton[31] has proposed the use of a 3-m (10-ft) deep anaerobic cell preceding either hyacinth or duckweed covered basins. Odor control in the anaerobic cell is provided by a surface mat of duckweed plants 1 to 2 cm (0.4 to 0.8 in) thick.

Since the duckweed plants do not play a major role in direct pollutant removal the use of inlet diffusers for initial wastewater distribution is not critical, as it was for hyacinth systems. Nor is the maintenance of an aerobic zone for mosquito control a factor since the mosquito larvae will not be able to penetrate a fully developed duckweed mat and therefore are not a problem. An effluent manifold is desirable to ensure utilization of the entire basin width for treatment. A screen or other baffling system is essential at the outlet of the basin to prevent loss of the small floating plants with the effluent. The basin configurations suggested for duckweed ponds are $L:W = 15:1$ to ensure plug flow conditions.

Pond systems in colder climates could be designed for the seasonal use of duckweed to significantly improve performance during the normal algal growth season. The pond cells could be seeded with duckweed soon after all ice had melted, and their rapid growth should ensure high quality effluent for the balance of the summer. The mat of floating plants on these systems should be harvested prior to the onset of freezing weather.

Operation and maintenance. The major operational concerns with duckweed basins are essentially the same as for the facultative ponds described in Chap. 4. Maintenance and harvest of the vegetation, control of odors and mosquitos, and sludge removal and disposal/utilization require some special concern.

The potential for excessive evapotranspiration losses from duckweed basins is not as significant as the losses from hyacinths. This is due to the very small size of the individual plants and to the thick surface mat that can develop.

Mosquito and odor control. As long as a thick surface mat is maintained on the duckweed basins, mosquitos should not be a problem. The mosquito larvae cannot survive in the anaerobic water beneath the surface cover and cannot penetrate a thick mat to obtain oxygen. To ensure rapid regrowth and maintenance of odor control not more than 20 percent of a basin should be harvested at any one time during the plant's growing season.

Effluent odor control may be a concern because the water in the basin is likely to be anaerobic at all times. Post aeration may be necessary in some situations. Odors may also be a concern at seasonal duckweed systems in colder climates. These basins may experience a spring and fall "overturn" due to temperature induced density differences in the water column. The resuspension of benthic material during the overturn period can result in objectionable odors. The problem is not unique to duckweed basins but can occur with any type of unaerated pond in colder climates. The typical solution is to locate these ponds at least 0.4 km (0.25 mi) from any habitation.

Vegetation management. Frequent plant harvests are not usually necessary since the major function of the duckweed plant is to provide a cover on the water surface. Harvest schedules and rates have not yet been developed from the limited experience with duckweed systems. It is necessary to have an effective surface mat of plants but it would also seem desirable to conduct a regular partial harvest to encourage semivigorous growth and to remove dead and decaying plants. Harvesting a broad area is not recommended since the remaining plants will be subjected to wind-induced drift resulting in their lateral compaction and further loss of surface cover. A reasonable value might be 20 percent or less of the basin area at any one time. Harvesting procedures typically utilize some type of floating device such as booms or pusher boats to bring the plants to the bank of the treatment cell and then they are removed using a method similar to the one described for hyacinths. In small basins this final removal can be a manual operation due to the small size of the individual plants. At several operational systems in Minnesota a patented floating plastic grid contains the duckweed plants in hexagon shaped cells and the floating harvester rides over the top of the cells to collect the plants.

Sludge removal. Assuming that regular harvesting is practiced, the benthic sludge in these duckweed basins should be similar to sludges in conventional facultative treatment ponds; the cleaning procedures and frequency will also be similar. Chapter 4 and Refs. 18 and 29 provide guidance on sludge removal from pond systems.

Utilization/disposal of harvested plants. The harvested plants can be used directly in the wet state as animal feed if transportation requirements are minimal. If significant off-site transport is necessary then on-site air drying is recommended. Drying times and procedures should be similar to those described in the previous section on hyacinths. Composting of the harvested duckweed plants should also be feasible.

5.2 Submerged Plants

The use of submerged aquatic macrophytes for treatment of wastewater has been tested in the laboratory and greenhouse and in a pilot scale field study in Michigan.[17] Table 5.12 provides information on some of the submerged freshwater plants that have been studied or considered for use in wastewater treatment.

The desirable water temperature for these plants ranges from 10 to 25°C (50 to 77°F), with growing being inhibited at temperatures above

TABLE 5.12 Submerged Aquatic Plants with Potential for Wastewater Treatment

Common name, *scientific name*	Distribution	Characteristics
Pondweed, *Potamogeton* sp. *P. amplifolius* is most studied type.	Worldwide	Has both floating and submerged leaves, reproduces from rhizomes growing in sediments.
Water milfoil, *Myriophyllum heterophyllum*	Worldwide	Highly branched stem up to 3 m long, vegetative reproduction.
Water weed, *Elodea* sp. *E. canadensis* is most studied type	Cooler parts of North and South America	Irregular branching stem, vegetative reproduction.
Coontail, *Ceratophyllum demersum*	Throughout U.S.	Rootless, branched stem, pinnate leaves, vegetative reproduction.
Fanwort, *Cabomba caroliniana*	Tropical and temperate U.S.	Highly branched from the base, whorled leaves, vegetative reproduction.

35°C (95°F). Turbidity of the water must not be high enough to prevent light transmission to the plants to support their photosynthetic activity and an aerobic environment is also necessary.

Performance expectations

The environmental requirements discussed previously would suggest that submerged plants might be best suited for final nutrient removal from previously treated and clarified wastewater, but there have been small scale greenhouse tests using primary effluent in aerated containers.[9] The units with *Elodea nuttalli* did demonstrate significant removal of BOD, phosphorus, and nitrogen, but the performance was just slightly better than the control units that contained no plants. The other plant species that were tested (*Myriophyllum heterophyllum, Ceratophyllum demersum*) were rapidly fouled with filamentous algae which in turn reduced productivity and system performance. *Elodea* was also a component in the pilot scale pond systems tested in Michigan.[17] Very significant nitrogen and phosphorus removals (nitrate from 15 mg/L to 0.01 mg/L and phosphorus from 4 mg/L to 0.03 mg/L) were achieved, but were ascribed to factors other than plant uptake.

Design considerations

There are insufficient data for the development of process design criteria for pond systems based on submerged plants as a major treatment component. They may be suited for final effluent polishing after wetlands or other pond units. Full-scale units would require a shallow depth to ensure adequate light penetration and contact between the plants and the wastewater. Unfortunately, this same environment might also be an ideal setting for algae development requiring another process step for algal separation.

5.3 Aquatic Animals

The aquatic animals that have been considered for use in wastewater treatment include *Daphnia*, brine shrimp, and a wide variety of fish, clams, oysters, and lobsters in both monoculture and polyculture systems.[3,7,17,24] Except for the predatory fish and the lobsters, the primary function of the other species is the removal of suspended solids or algae. Assuming the animals are routinely harvested, this will in turn also improve nutrient removal.

Daphnia and brine shrimp

Daphnia are small crustaceans (1 to 3 mm in length). They are filter feeders, and the major direct contribution to wastewater treatment is the removal of suspended solids. A 10-day detention time *Daphnia* culture pond in Giddings, Texas averaged about 77 percent BOD removal (average influent BOD was 54 mg/L) over a 2-month test period.[7] When cultured in wastewater the *Daphnia* is very sensitive to pH, since high pH values will permit the presence of un-ionized ammonia (NH_3) which is toxic to the animal. A functional system would therefore require shading to suppress algae, which if uncontrolled would result in elevated pH levels. In some cases gentle aeration and the addition of supplemental acid may be necessary. When all of these management requirements are considered the use of *Daphnia* culture basins for waste treatment is probably not cost effective.

Brine shrimp require saline waters for survival and that limits the applicable range for this organism. They have been used in laboratory and pilot scale experiments in a two-step process where the brine shrimp are expected to clarify the effluent from algae ponds. Removal of BOD and SS averaged 89 percent in pilot scale shrimp tanks that were aerated and heated during the winter months.[7] The environmental and management requirements for this shrimp culture may not permit cost-effective full-scale systems.

Fish

The utilization of fish has involved their placement in wastewater treatment ponds, the addition of wastewater effluent to fish ponds, and the sequential conversion of wastewater nutrients to algae and microinvertebrates prior to discharge to fish ponds. Table 5.13 lists the fish species which have been used in wastewater treatment systems.

Fish activity is highly dependent on temperature and most of the species listed in Table 5.13, with the exception of catfish and minnows, require relatively warm water. Oxygen levels are also a critical parameter, dissolved oxygen concentrations less than 2 mg/L are limiting and concentrations below 5 mg/L will allow only slow growth. The presence of un-ionized ammonia (NH_3) is also toxic to the young of the larger species. Sources of this ammonia are both the wastewater and the wastes from the fish population. The combination of these factors limits the use of the larger fish species to the final cells of treatment systems or the application of well-oxidized wastewater to fish ponds.

TABLE 5.13 Fish Species Used in Wastewater Treatment

Common name, *scientific name*	Pond location	Feeding habits
Silver carp, *Hypophthalmichthys molitrix*	Upper layers	Phytoplankton
Bighead carp, *Aristichthys nobilis*	Upper layers	Phytoplankton, zooplankton suspended solids
Black carp, *Mylopharyngodon piceus*	Bottom	Snails, crustaceans, mussels
Grass carp, *Ctenopharyngodon idella*	Ubiquitous	Variable
Common carp, *Cyprinis carpio*	Bottom	Phytoplankton, zooplankton, insect larvae
Tilapia, *Tilapia* spp., *Sarotherodon* spp.	Ubiquitous	Plants, plankton, detritus, invertebrates
Catfish, *Ictalurus* spp.	Bottom	Crustaceans, algae, fish insect larvae
Fathead minnows, *Pinephales promelas* Golden shiner, *Notemigonas crysoleucas*	Bottom	Phytoplankton, zooplankton invertebrates
Mosquito fish, *Gambusia affinis*	Surface	Insect larvae, zooplankton, algae
Buffalofish, *Ictiobus* spp.	Bottom	Crustaceans, detritus, insect larvae

Other parameters of concern are salinity, metals, and toxic substances. The pH tolerance of fish ranges from about 6.5 to 9 and this will not usually be a constraint in most wastewater systems.

Performance expectations. There have been two major studies to evaluate treatment performance in wastewater treatment ponds that include fish. At the Quail Creek wastewater pond system channel catfish were stocked in the third and fourth cells, fathead minnows and *Talapia* were stocked in the third cell, and golden shiner minnows were placed in the fifth and sixth cells.[3] The first two cells of this 140-day detention time system were aerated. The organic loading on the entire system was about 47 kg/(ha · day) [42 lb/(acre · day)], and the organic loading on the first cell containing fish (number 3) was about 34 kg/(ha · day) [31 lb/(acre · day)]. The detention time in the final four cells containing fish was close to 90 days. The BOD was reduced in these units from 24 mg/L to 6 mg/L and the SS from 71 mg/L to 12 mg/L. The initial stocking rate for the fish was 29 kg/ha (26 lb/acre), the fish biomass increased significantly, and the net production was 44 kg/(ha · month) [39 lb/(acre · month)] but there is no direct evidence that the fish contributed significantly to wastewater treatment.

The study at the Benton Services Center, Arkansas also utilized a six-cell pond system, but without aeration.[11] The first study phase compared parallel operation of three cells in series, with one set stocked with silver, grass, and bighead carp and the other set operated as a control without fish. Performance of the two sets was generally similar but the fish culture units showed somewhat better performance. The effluent from the fish unit had BOD ranging from 7 to 45 mg/L with values less than 15 mg/L obtained more than 50 percent of the time. The control set had BOD ranging from 12 to 52 mg/L with values less than 23 mg/L about 50 percent of the time. Effluent suspended solids were similar for the two systems most of the time.

The second phase of the study used all six cells in series with a new baffle constructed in each to reduce short circuiting of flow. Silver carp and bighead carp were stocked in the last four cells and additional grass carp, buffalo fish, and channel catfish in the final cell. No supplemental feed or nutrients were added to the fish culture ponds. The initial stocking rate was 426 kg/ha (380 lb/acre) and the estimated net production during the 8-month study was 417 kg/(ha · month) [390 lb/(acre · month)]. The BOD removal for the entire six-cell system averaged 96 percent with about 89 percent of the removal occurring in the first two conventional stabilization cells. Overall performance was better than the original unmodified six-cell system but it is not clear whether the improvement was due to the fish or the improved flow pattern following construction of the baffles. It is likely that the fish did contribute to the low suspended solids in the final effluent (17 mg/L) via algae predation. These two examples, plus successful experience elsewhere raising catfish, fathead minnows, *Tilapia*, rainbow trout, and muskellung[13] in treated wastewater confirm the feasibility of using wastewater in fish culture. Unfortunately there is little evidence that fish culture can provide significant cost-effective benefits for wastewater treatment. The final lightly loaded cells in wastewater pond systems can be used for fish culture if a market for the harvested fish exists. At present, federal and state health regulations would prevent the sale of such fish for direct human consumption even though microbiological studies have not detected any contamination. The major markets for this harvested material would be bait fish, pet food, or fertilizer.

Marine polyculture

Several systems have been proposed or tested using a variety of marine animals and plants for polishing of wastewater effluents. The principal goal is usually nitrogen removal and the first step is to convert wastewater nitrogen to biomass in algae ponds and then use marine organisms with a commercial value to consume the algae. A pilot system of this type was constructed at the Woods Hole Oceanographic Institute. Secondary effluent was diluted with seawater and then introduced into shallow algae ponds. It was then sent through aerated channels with stacked trays containing American oysters, hard clams, and lobsters and finally into an aerated cell for seaweed production.[32] It was not possible to control the seasonal variation in algal species, which resulted in slow growth rates and high mortalities of the shellfish culture. The overall cost effectiveness of the concept is questionable. Success will depend on the ability to limit algal growth in the initial ponds to desirable species and this may not be practical or possible in a full-scale system in the field.

REFERENCES

1. Amasek, Inc.: *Assessment of Operations, Water Hyacinth Nutrient Removal Treatment Process Pilot Plant,* Florida Dept. of Environmental Regulation, FEID 59-6000348, Kissimmee, Fla., 1986.
2. Babbitt, H. E., and E. R. Baumann: *Sewerage and Sewage Treatment,* John Wiley & Sons, N.Y., 1952.
3. Coleman, M. S.: "Aquaculture as a Means to Achieve Effluent Standards," *Proceedings Wastewater Use in the Production of Food and Fiber,* EPA 660/2-74-041, EPA, Washington, D.C., 1974, pp. 199–214.
4. Conn, W. M., and A. C. Langworthy: "Practical Operation of a Small Scale Aquaculture," *Proceedings Water Reuse III,* American Water Works Association, Denver, Colo., 1985, pp. 703–712.
5. DeBusk, T. A.: *Community Waste Research at the Walt Disney World Resort Complex,* Reedy Creek Utilities Co., Lake Buena Vista, Fla., 1986.

6. Dinges, R.: "Development of Hyacinth Wastewater Treatment Systems in Texas," *Proceedings Aquaculture Systems for Wastewater Treatment,* EPA 430/9-80-006, MCD 67, EPA Office of Municipal Pollution Control, Washington, D.C., 1979, pp. 193–231.
7. Dinges, R.: *Natural Systems for Water Pollution Control,* Van Nostrand Reinhold, New York, 1982.
8. Doersam, J.: "Use of Water Hyacinths for the Polishing of Secondary Effluent at the City of Austin Hyacinth Greenhouse Facility," *Proceedings of Conference on Aquatic Plants for Water Treatment and Resource Recovery,* Orlando, Florida, July 1986, University of Florida, July 1987.
9. Eighmy, T. T., and P. L. Bishop: *Preliminary Evaluation of Submerged Aquatic Macrophytes in a Pilot-Scale Aquatic Treatment System,* Dept. of Civil Engineering, University of New Hampshire, Durham, N.H., 1985.
10. Gee & Johnson Engineers: *Water Hyacinth Wastewater Treatment Design Manual for NASA/NSTL,* West Palm Beach, Fla., 1980.
11. Henderson, S.: "Utilization of Silver and Bighead Carp for Water Quality Improvement," *Proceedings Aquaculture Systems for Wastewater Treatment,* EPA 430/9-80-006, Environmental Protection Agency, Washington, D.C., 1979, pp. 309–350.
12. Hillman, W. S., and D. C. Culley: "The Use of Duckweed," *American Scientist,* 66:442–451, 1978.
13. Hyde, H. C., R. S. Ross, and L. Sturmer: *Technology Assessment of Aquaculture Systems for Municipal Wastewater Treatment,* EPA 600/2-84-145, Environmental Protection Agency Municipal Engineering Research Laboratory, Cincinnati, Ohio, 1984.
14. Kamber, D. M.: *Benefits and Implementation Potential of Wastewater Aquaculture,* Contract Rpt. 68-01-6232, EPA Office of Regulations and Standards, January 1982.
15. Lee, C. L., and T. McKim: *Water Hyacinth Wastewater Treatment System,* Reedy Creek Utilities Co., Buena Vista, Fla., 1981.
16. Leslie, M.: *Water Hyacinth Wastewater Treatment Systems: Opportunities and Constraints in Cooler Climates,* EPA 600/2-83-075, EPA, Washington, D.C., 1983.
17. McNabb, C. D.: "The Potential of Submerged Vascular Plants for Reclamation of Wastewater in Temperate Zone Ponds," *Biological Control of Water Pollution,* University of Pennsylvania Press, Philadelphia, Pa., 1976, pp. 123–132.
18. Middlebrooks, E. J., C. H. Middlebrooks, J. H. Reynolds, G. Z. Watters, S. C. Reed, and D. B. George: *Wastewater Stabilization Lagoon Design, Performance, and Upgrading,* Macmillan, New York, 1982.
19. Penfound, W. T., and T. T. Earle: "The Biology of the Water Hyacinth," *Ecological Monographs,* 18(4):447–472, 1948.
20. Reddy, K. R.: "Nutrient Transformations in Aquatic Macrophyte Filters Used for Water Purification," *Proceedings Water Reuse III,* American Water Works Association, Denver, Colo., 1985, pp. 660–678.
21. Reddy, K. R., and W. F. DeBusk: "Nutrient Removal Potential of Selected Aquatic Macrophytes," *J. Environ. Qual.,* 14(4):459–462, 1985.
22. Reddy, K. R., and D. L. Sutton: "Water Hyacinths for Water Quality Improvement and Biomass Production," *J. Envirn. Qual.,* 13(1):1–8, 1984.
23. Reed, S. C., R. Bastian, and W. Jewell: "Engineers Assess Aquaculture Systems for Wastewater Treatment," *Civil Engineering,* July 1981, pp. 64–67.
24. Ryther, J. H.: "Treated Sewage Effluent as a Nutrient Source for Marine Polyculture," *Proceedings Aquaculture Systems for Wastewater Treatment,* EPA 430/9-80-006, EPA, Washington, D.C., 1979, pp. 351–376.
25. Stephenson, M., G. Turner, P. Pope, J. Colt, A. Knight, and G. Tchobanoglous: *The Use and Potential of Aquatic Species for Wastewater Treatment,* "Appendix A The Environmental Requirements of Aquatic Plants," Publication No. 65, California State Water Resources Control Board, Sacramento, Calif., 1980.
26. Stewart, E. A.: "Utilization of Water Hyacinths for Control of Nutrients in Domestic Wastewater—Lakeland, Florida," *Proceedings Aquaculture Systems for Wastewater Treatment,* EPA 430/9-80-006, MCD 67, EPA Office of Municipal Pollution Control, Washington, D.C., 1979, pp. 273–293.

27. Swett, D.: "A Water Hyacinth Advanced Wastewater Treatment System," *Proceedings Aquaculture Systems for Wastewater Treatment,* EPA 430/9-80-006, MCD 67, EPA Office of Municipal Pollution Control, Washington, D.C., 1979, pp. 233–255.
28. Tchobanoglous, G.: Personal Communication.
29. US EPA. *Design Manual Municipal Wastewater Stabilization Ponds,* EPA 625/1-83-015, EPA Center for Environmental Research Information, Cincinnati, Ohio, 1983.
30. Weber, A. S., and G. Tchobanoglous: "Nitrification in Water Hyacinth Treatment Systems," *J. Environ. Eng. Div.,* ASCE, 11(5):699–713, 1985.
31. Wolverton, B. C.: "Engineering Design Data for Small Vascular Aquatic Plant Wastewater Treatment Systems," *Proceedings Aquaculture Systems for Wastewater Treatment,* EPA 430/9-80-006, MCD 67, EPA Office of Municipal Pollution Control, Washington, D.C., 1979, pp. 179–192.
32. Wolverton, B. C., and R. C. McDonald: "Nutritional Composition of Water Hyacinths Grown on Domestic Sewage," *Economic Botany,* 32(4):363–370, 1978.
33. Wolverton, B. C., and R. C. McDonald: "Upgrading Facultative Wastewater Lagoons with Vascular Aquatic Plants," *J. Water Pollution Control Fed.,* 51(2):305–313, 1979.

Chapter

6

Wetland Systems

Wetlands are defined in this book as land where the water surface is near the ground surface long enough each year to maintain saturated soil conditions, along with the related vegetation. These wetland systems are typically characterized by emergent aquatic vegetation such as cattails (*Typha*), rushes (*Scirpus*), and reeds (*Phragmites*). They can also contain some of the floating and submerged plant species discussed in Chap. 5 as well as phraetophytes (plants rooted on the bank but growing out over the water). Willows and other low vegetation are common to many bogs and peatlands in cooler climates, while cypress, ash, willow, and other trees are often found in southern swamps and strands. Performance expectations for the major wetland types of wastewater treatment are summarized in Table 1.2 in Chap. 1. Four major combinations involving wastewater and wetlands can be observed in the United States.[14]

- Disposal of treated effluent into natural wetlands
- Use of natural wetlands for further wastewater renovation
- Use of effluents or partially treated wastewater for enhancement, restoration, or creation of wetlands
- Use of constructed wetlands as a wastewater treatment process

All four categories provide some degree of wastewater treatment, either directly or indirectly. In the United States, however, there are some constraints on the use of many natural wetlands as functional components of wastewater treatment systems. These wetlands are considered under the law to be "waters of the United States," and as such a permit is required for any discharge. The water quality requirements for this discharge are specified by the applicable federal, state, and/or local agencies and are typically at least equal to secondary effluent standards.

Most states (except for Florida, and a few others considering special wetland standards) make no distinction between the wetland and the adjacent surface waters and apply the same requirements to both. Under these conditions economics will not favor the utilization of natural wetlands as a major component in a wastewater treatment process since the basic treatment must be provided prior to discharge to the wetland. The economics are further constrained in that the purchase of wetlands for this purpose is not eligible for federal construction grant funding, as is the purchase of land for construction of the land treatment systems described in Chap. 7. Special situations may arise in which natural wetlands may provide further effluent polishing or, if the wetland is isolated from other surface waters, more basic treatment. Such projects may still have to overcome public concerns regarding the wetland habitat since introduction of significant quantities of treated wastewater is likely to alter conditions.

The use of treated effluent for enhancement, restoration, or creation of wetlands can be a very desirable and environmentally compatible activity. In one example stabilization pond effluent has been used to increase the marsh habitat for the endangered Mississippi sandhill crane.[12] Descriptions of similar activities can be found in Ref. 3, 7, and 13.) In these cases the water and the contained nutrients are beneficially utilized, but the projects are not typically intended for wastewater treatment.

The wetland concept most likely to offer a cost-effective treatment alternative is the use of constructed wetlands. Constructing a wetland where one did not exist before avoids the regulatory entanglements associated with natural wetlands and allows design of the wetland for optimum wastewater treatment. Typically, a constructed wetland should perform better than a natural wetland of equal area since the bottom is usually graded and the hydraulic regime in the system is controlled. Process reliability is also improved because the vegetation and the other system components can be managed as required. The constructed wetland is therefore the major focus of this chapter. However, much of the information provided on plant responses, treatment reactions, and other factors is also applicable to natural wetlands.

The constructed wetlands described in this chapter include the conventional wetland with an exposed free water surface, as well as the more recently developed vegetated submerged bed (VSB) concepts. The latter method involves subsurface flow through a permeable medium. Both concepts utilize emergent aquatic vegetation, and as described in Sec. 6.1, they all depend on the same basic microbiological reactions for treatment.

In addition to municipal wastewaters, constructed wetlands have been used for a variety of industrial applications. The *free-water-surface wetland* is widely used as an inexpensive method of treating acid mine drainage. Over 20 such systems were constructed in 1984–1985 in Pennsylvania, West Virginia, Ohio, and Maryland. The wetland exposes the acid mine wastes to aerobic oxidizing conditions prior to discharge to the receiving stream. Iron concentrations of 25 to 100 mg/L can be reduced to less than 2 mg/L. The pH is also modulated, and sometimes limestone riprap is also used to raise the pH to 6 or higher. Constructed wetlands for industrial wastewaters with a very high organic content require a preliminary step for the partial removal of these materials.

6.1 Wetland Components

The major system components having some influence in the treatment process in wetlands includes the plants, soils, bacteria, and animals. Their function and the system performance are in turn influenced by water depth, temperature, pH, and dissolved oxygen concentration.

Plants

A wide variety of aquatic plants have been used in wetland systems designed for wastewater treatment. The larger trees (cypress, ash, willow, etc.) preexist on isolated natural bogs, strands, and "domes" used for wastewater treatment in Florida and elsewhere. There has been no attempt to use these species in a constructed wetland, nor has their function as a treatment component in the system been defined.

The emergent plants most frequently found in wastewater wetlands include cattails, reeds, rushes, bulrushes and sedges. Table 6.1 provides information on their distribution in the United States and some of the major environmental requirements of each.

Influence of water level. The water level in the system and the duration of flooding can be important factors for the selection and maintenance of wetland vegetation. Cattails grow well in submerged soils and may dominate where standing water depth is over 15 cm (6 in). Reeds occur along the shorelines of water bodies where the water table is below the surface but will also grow in water deeper than 1.5 m (5 ft). Growth is best in standing water but the depth seems to have no direct effect.

TABLE 6.1 Emergent Aquatic Plants for Wastewater Treatment[16]

Common name, *scientific name*	Distribution	Temperature, °C		Maximum salinity tolerance, ppt†	Optimum pH
		Desirable	Survival*		
Cattail, *Typha* spp.	Throughout the world	10–30	12–24	30	4–10
Common reed, *Phragmites communis*	Throughout the world	12–33	10–30	45	2–8
Rush, *Juncus* spp.	Throughout the world	16–26	—	20	5–7.5
Bulrush, *Scirpus* spp.	Throughout the world	16–27	—	20	4–9
Sedge, *Carex* spp.	Throughout the world	14–32	—	—	5–7.5

* Temperature range for seed germination; roots and rhizomes can survive in frozen soils.
† ppt = parts per thousand.

The common reed is a poor competitor and may give way to other species in nutrient-rich shallow waters. Bulrushes can tolerate long periods of soil submergence and occur at water depths of 0.75 to 25 cm (0.3 to 10 in) in California.[16] In deeper water bulrushes may give way to cattails. Sedges generally occur along the shore or in shallower water than bulrushes.

Evapotranspiration losses. The water losses due to evapotranspiration (ET) can affect the feasibility of the various wetland designs in arid climates and their performance during the warm summer months in all locations. In the western states, where appropriate laws govern the use of water, it may be necessary to replace the volume of water lost to protect the rights of downstream water users. Evaporative water losses in the summer months decrease the water volume in the system, and therefore the concentration of remaining pollutants tends to increase even though treatment is very effective on a mass removal basis. The decreased water volume also increases the detention time and may increase the potential for anoxic or anaerobic conditions; this will affect performance and may increase the risk of mosquito development in wetlands with an exposed water surface.

The unit transpiration rate for cattails is low, but the high mass of leaves results in a high overall rate. A rate of 1.3 to 1.6 times the rate from open water was observed for cattails in India, while culture tanks with cattails in Poland showed an increase of 2.75 to 3.33 times the open-water rate. The unit transpiration rate for reeds is high, but the low unit mass of leaves moderates the water losses. The ET losses from culture tanks in Poland were reported to be about 2.1 to 2.3 times the evaporation from open water. The ET losses for bulrushes in culture tanks in Poland was reported to be 3.2 to 3.5 times the open-water rate.[16] The ET losses observed for sedges in India was about 2.4 times the losses from adjacent open water.[16] The ET loss measurements from small culture tanks tend to overestimate the actual losses owing to the small size of the containers. As a result, the culture tank values reported above can probably be reduced by at least 50 percent for application to a full-scale wetland treatment system.

Oxygen transfer. If wetland soils are submerged for long periods, they will almost certainly be anaerobic. All the emergent plants listed in Table 6.1 have the capability of absorbing oxygen and other needed gases from the atmosphere through their leaves and above-water stems, and they have large gas vessels, which conduct those gases to the roots. As a result the soil zone in immediate contact with the roots or rhizomes (rootstalks) can be aerobic in an otherwise anaerobic environment. It has been estimated that these plants can transfer 5 to

45 g oxygen per day per square meter of wetland surface area depending on plant density and oxygen stress levels in the soil.[1,10] Some of this oxygen will support aerobic microbes in and around the root system. Systems utilizing reeds may be more effective in this oxygen transfer because the rhizomes penetrate vertically, and more deeply than cattails. Reeds have been the preferred plant species for the submerged-bed systems described in Sec. 6.3 owing to this greater penetration.

Plant diversity. Natural wetlands typically contain a wide diversity of plant life. There is however, no need to attempt to reproduce the natural diversity in a constructed, free-water-surface wetland designed for optimum wastewater treatment. Such attempts in the past have shown that eventually cattails alone or in combination with either reeds or bulrushes will dominate in a wastewater system owing to the high nutrient levels.

Plant functions. As described in Sec. 6.2, the plants can contribute to treatment through uptake of nutrients and other wastewater constituents. Their major role in addition to transferring oxygen to the root zone is due to their physical presence in the system: the stalks, roots, and rhizomes penetrate the soil or support medium, allowing more effective fluid movement and contact opportunities in the benthic layer. The stalks and leaves above the water surface provide a shading canopy, which limits sunlight penetration and therefore controls algae growth. Perhaps most important in the free-water-surface wetlands are the submerged portions of the leaves, stalks, and litter, which serve as the substrate for attached microbial growth. It is the responses of this attached biota that is believed responsible for much of the treatment that occurs.

Soils

Except for carbon, most of the nutrients required for plant growth are obtained from the soil by emergent aquatic plants. The soil in natural cattail stands has been described as muck, but the plants will grow in a wide variety of media. Similarly, the soil in natural stands of reeds may be clay, sand, or silt, but the plant can be propagated in many different soils and even in gravel, although in natural habitats the plants are seldom found on gravel beds. The bulrush has capabilities similar to cattails, while sedges have been grown in silty clays and typical garden soils.

The void spaces in the soil or other media serve as the flow channels for the VSB wetlands. Treatment in these cases is provided by microbial organisms attached to the roots and rhizomes and the adjacent soil surfaces. Soils may be selected as the bed media when phosphorus removal is a project requirement. As described in Chaps. 3 and 7, phosphorus removal in such a soil matrix can be very effective for many decades. However, soils with high phosphorus removal potential tend to be finer in texture, with low to moderate permeability, and this may limit the hydraulic capacity of a wetland soil bed.

Organisms

A wide variety of beneficial organisms, ranging from bacteria to protozoa to higher animals, can exist in wetland systems. The range of species present is similar to that found in the aquaculture and pond systems described in previous chapters. Wetland systems are similar to the hyacinth systems described in Chap. 5 in that attached microbial growth is believed to be the major contributor to wastewater treatment. In the case of hyacinths the growth occurs on the floating roots; in the case of emergent vegetation in wetlands the growth occurs on the submerged portions of the plant, on the litter and other detritus, or directly on the medium for the submerged VSB types.

In effect, both types of wetlands and the overland flow (OF) concept described in Chap. 7 are all "attached-growth" biological systems and share many similarities with the familiar trickling filters and rotating biological contactors (RBCs). All these systems require a substrate for the development of the biological growth; their performance is dependent on the detention time in the system and on the contact opportunities provided and is regulated by the availability of oxygen and the temperature. Section 6.3 describes design procedures for all these natural attached growth processes.

6.2 Performance Expectations

Wetland systems can reduce high levels of biochemical oxygen demand (BOD), suspended solids (SS), and nitrogen, as well as significant levels of metals, trace organics, and pathogens. If appropriate soils are utilized, VSB systems can also remove high levels of phosphorus. The basic treatment mechanisms are similar to those described in Chaps. 4 and 5 and include sedimentation, chemical precipitation and adsorption, and microbial interactions with BOD, SS, and nitrogen, as well as some uptake by the vegetation. The performance of several pilot-scale wetland systems is summarized in Table 6.2.

BOD removal

The removal of settleable organics is very rapid in all wetland systems and is due to the quiescent conditions in the free-water-surface types

TABLE 6.2 Performance of Pilot-Scale Constructed Wetland Systems

		Effluent concentration, mg/L					
Location	Wetland type	BOD	SS	NH_3	NO_3	Total N	Total P
Listowel, Ont.[8]	Open water, channel	10	8	6	0.2	8.9	0.6*
Arcata, Calif.[5]	Open water, channel	< 20	< 8	< 10	0.7	11.6	6.1
Santee, Calif.[6]	Gravel-filled channels	< 30	< 8	< 5	< 0.2	—	—
Vermontville, Mich.[22]	Seepage basin wetland	—	—	2	1.2	6.2	2.1

* Alum treatment provided prior to the wetland component.

and to deposition and filtration in the VSB systems. Similar results have been observed with the OF systems described in Chap. 7, where close to 50 percent of the applied BOD is removed in the first few meters of travel down the treatment slope.

In free-water-surface wetlands, removal of the soluble BOD is due to the same factors described in Chap. 4, with the major contribution from the attached microbial growth. The major source of oxygen for these reactions is reaeration at the water surface, since algae are typically not present. Any excess oxygen transmitted by the plant to the root zone is likely to be consumed in the soil profile and not to contribute significantly to oxygen levels in the water. Wind-induced water turbulence and mixing will also be reduced or eliminated if a dense stand of vegetation is present. In northern climates, which experience freezing conditions of long duration, the development of a significant ice cover on the wetland system is likely. If the ice persists for more than a few days, depletion of oxygen in the system is possible, since surface reaeration is prevented. A final concern involves summer operation, during which high ET rates occur. These water losses can result in sluggish flow and an extended detention time, which may lead to anoxic or anaerobic conditions.

All these restrictions on the oxygen supply, then, impose limits on system design and operation. Specific criteria are presented in later sections, but in essence wetlands of this type are only suitable for low to moderate organic loadings. The organic loading should be distributed over a significant portion of the area and not applied at a single point. The design water depth should be 60 cm (24 in) or less to ensure adequate oxygen distribution, and partial effluent recirculation might be considered in the summer months to overcome ET losses and maintain design flow rates and oxygen levels. Supplemental mechanical aeration will probably not be cost-effective owing to the shallow water depths.

The major oxygen source for the subsurface components (soil, gravel, rock, and other media, in trenches or beds) is the gases transmitted by the vegetation to the root zone. In most cases the system is designed to maintain flow below the surface of the bed, so there can be very little direct atmospheric reaeration. The selection of plant species can therefore be an important factor. Work at the pilot wetlands in Santee, California[6] indicated that most of the horizontally growing root mass of cattails was confined to the top 30 cm (12 in) of the profile. The root zone of reeds extended to more than 60 cm (24 in) and bulrushes to 76 cm (30 in). In the cooler climate of western Europe the effective root zone depth for reeds is also considered to be 60 cm (24 in).[1] The gravel bed at the Santee system was 76 cm (30 in) deep, and the water level was maintained just below the surface.[6] The BOD removals observed (see Table 6.5) in the three parallel bulrush, reed, and cattail units reflect the expanded aerobic zone made possible by a deeper root penetration.

Suspended solids removal

Suspended solids removal is very effective in both types of constructed wetlands, as shown by the data in Table 6.2. Most of the removal occurs within the first few meters beyond the inlet, owing to the quiescent conditions and the shallow depth of liquid in the system. At the pilot system in Arcata, California essentially all the solids removal occurred in the initial 12 to 20 percent of the cell area.[5] At the Listowel, Ontario system some of the channels received unsettled effluent from an aeration cell, and SS reached 406 mg/L at times.[8] Removal was still very effective but resulted in the build-up of a sludge bank and dieback of cattails near the inlet. The cattails at the gravel bed wetland in Santee, California also experienced dieback following extended (more than 6 months) application of primary effluent; the bulrushes and reeds in this system survived the same conditions. The dieback is apparently due to oxygen stress caused by the shallow root penetration of the cattail plant.

Nitrogen removal

Nitrogen removal is very effective in both the free-water-surface and submerged-flow constructed wetlands, and the major removal mechanisms are similar for both. Although plant uptake of nitrogen does occur, only a minor fraction of the total nitrogen can be removed in these systems. At the Listowel system, where plant harvest was practiced, the harvested material accounted for less than 10 percent of the nitrogen removed by the system.[8] A range of 12 to 16 percent was estimated for removal by plant uptake at the Santee, California, system.[6] Total nitrogen removals of up to 79 percent are reported at nitrogen loading rates (based on elemental N) up to 44 kg/(ha · day) [39 lb/(acre · day), in a variety of wetland systems.[22]

The major contribution to nitrogen removal, as with the hyacinth systems described in Chap. 5, is believed to result from nitrification/denitrification. The opportunity for nitrification exists within the aerobic attached growth biota, and there are adjacent anaerobic sites and carbon sources in the litter and benthic layer. The penetration of the plant roots and rhizomes into the media in subsurface flow systems provides the aerobic microsites for nitrification in an otherwise anaerobic environment. The depth of the bed and the type of plant used can make a significant difference. At Santee, with the same bed depth for all channels, the bulrush system removed 94 percent of the applied nitrogen while the reeds, cattails, and unvegetated beds achieved 78, 28, and 11 percent removal, respectively.[6] This clearly demonstrates the necessity of vegetation as a component in the system and also the importance of matching the depth of the bed in subsurface systems to the potential root penetration depth for the selected vegetation. Any flow beneath the root zone in the subsurface media will be anaerobic and will not contribute significantly to removal of nitrogen or other constituents.

Nitrification was incomplete at times in the free-water-surface system at Listowel, Ontario, owing to occasional anoxic conditions in the water. In the summer months this was caused by sluggish flow (due to high ET losses) and high oxygen demand for organic decomposition. Correction of this situation should be possible with effluent recirculation, which is routinely practiced with other attached-growth biological processes.

At Listowel oxygen transfer to the water in winter was limited by the ice cover, so the low oxygen levels and low temperatures limited nitrification [liquid temperatures were about 3°C (37°F) in the winter]. Studies with trickling filter and RBC systems indicate that significant levels of nitrification by attached growth organisms require a temperature of 5 to 7°C (41 to 45°F).[15] This may be a significant process limitation for exposed-water-surface wetland systems in cold climates if year-round nitrification is a process expectation.

Phosphorus removal

Phosphorus removal in natural systems occurs as a result of adsorption, complexation, and precipitation and is very effective in the soil-based land treatment systems described in Chap. 7. Phosphorus removal in many wetland systems is not very effective because of the limited contact opportunities between the wastewater and the soil. The exceptions are the submerged bed designs when proper soils are selected as the medium for the system. A significant clay content and the presence of iron and aluminum will enhance the potential for phosphorus removal. A VSB system with the proper soils should perform as well and have a useful life for phosphorus removal as long as that of the rapid infiltration (RI) process described in Chap. 7 (see Sec. 3.6 for further discussion of phosphorus removal mechanisms).

Metals removal

There are limited data available on the metal removal capability of free-water-surface wetlands; since the removal mechanisms are similar to those described above for phosphorus, the response should be about the same. There is greater opportunity for contact and adsorption in the subsurface flow wetlands, and metals removal can be very effective. A VSB system, with soil medium, should perform as well as the RI process described in Chap. 7. Removal in gravel bed systems can also be very effective owing to adsorption on the media and contained organics (see Sec. 3.5 for further discussion and some results from the Santee, California pilot wetlands).

Pathogen removal

Pathogen removal in free-water-surface wetlands is due to the same factors described in Chap. 3 for pond systems, and Eq. 3.25 can be used to estimate pathogen removal in these wetlands. Table 3.9 contains performance data for both free-water-surface (Arcata, California) and subsurface-flow (Santee, California, Iselin, Pennsylvania) wetlands. The removal in the latter cases is related to the texture and grain size of the medium. Gravel was used in Santee, California and a mixture of gravel and sand at Iselin, Pennsylvania, and the finer-textured material is clearly superior. A VSB system using fine to moderately coarse soils should perform as well as the slow rate land treatment system described in Chap. 7 (see Sec. 3.4 in Chap. 3 for further discussion).

6.3 Design Procedures

All constructed wetland systems can be considered to be attached-growth biological reactors, and their performance can be described with first-order plug-flow kinetics. Design equations are presented in this section for both the free-water-surface and subsurface-flow (VSB) types. These equations were derived from a relatively limited data base, so caution should be used in their application. They should give a reasonable, and hopefully conservative, estimate of design requirements. A pilot test is strongly recommended for large-scale projects.

Constructed wetlands with a free water surface

These systems typically consist of basins or channels, with some sort of subsurface barrier to prevent seepage, soil or another suitable medium to support the emergent vegetation, and water at a relatively shallow depth flowing through the unit. The shallow water depth, low flow velocity, and presence of the plant stalks and litter regulate water flow and, especially in long, narrow channels, ensure plug-flow conditions. The conditions for biological treatment are similar to those for the OF concept described in Chap. 7 and to trickling filters and similar attached-growth systems described in Refs. 4, 5, and 18. The basic relationship for plug-flow reactors is given by Eq. 6.1.

$$\frac{C_e}{C_0} = \exp\left[-K_T t\right] \tag{6.1}$$

where C_e = effluent BOD, mg/L
C_0 = influent BOD, mg/L
K_T = temperature-dependent first-order reaction rate constant, days^{-1}
t = hydraulic residence time, days

The relationship takes the form shown in Eq. 6.2 for the OF concept described in Chap. 7. This equation can be derived by design procedures presented in Ref. 20. The constants and coefficients were developed from research in New Hampshire and validated with performance data collected elsewhere.[20]

$$\frac{C_e}{C_0} = A \exp\left(-\frac{0.078\, K_T L W}{S^{1/3} Q}\right) \tag{6.2}$$

where A = fraction of applied BOD that is not settleable in the first few meters of flow
= 0.52
K_T = 43.2, temperature-dependent rate constant, days^{-1}
L = length of flow path, m (ft)
W = width of treatment area, m (ft)
S = slope, or hydraulic gradient of flow system (as a decimal fraction)
Q = average flow in the system, m^3/day (ft^3/day)
$= \dfrac{Q_{\text{influent}} + Q_{\text{effluent}}}{2}$

The average flow is used to compensate for ET losses or precipitation additions. As a first approximation it can be assumed that Q is equal to the design hydraulic loading. (The other terms have been defined previously.)

Eckenfelder[4] has suggested that trickling filter systems can be described with first-order plug-flow kinetics and has offered Eqs. 6.3, 6.4, and 6.5 for this purpose.

$$\frac{C_e}{C_0} = \exp(-K_T x t) \tag{6.3}$$

where K_T = first-order reaction rate constant, days^{-1}
x = mass of active microbes in contact with the wastewater
= A_V, the specific surface of the medium available for development of attached-growth organisms, m^2/m^3 (ft^2/ft^3)
t = hydraulic residence time in the system
C_e = effluent BOD, mg/L
C_0 = influent BOD, mg/L

The hydraulic residence time t is given by:

$$t = \frac{CL}{(Q_a)^b} \tag{6.4}$$

where C and b are characteristics of the filter media
L = minimum length of flow path, i.e., filter depth, m (ft)
Q_a = hydraulic loading per unit surface area of the filter, m^3/(m^2 · day) [gal/(ft^2 · min)]

The C value is determined with Eq. 6.5

$$C = C'(A_V)^m \tag{6.5}$$

where C' and m are characteristics of the medium. The values for a wide variety of media were $C' = 0.7$ and $m = 0.75$.

Combining Eqs. 6.3, 6.4, and 6.5 produces:

$$\frac{C_e}{C_0} = \exp\left[-\frac{K_T C'(A_V)^{m+1} L}{(Q_a)^b}\right]$$

In an attached-growth wetland system Q_a is the average flow, or hydraulic loading, per unit cross-sectional area of the wetland (Q/dW), and since the contact surfaces of the medium are entirely submerged, the coefficient b should equal 1. Including these factors and combining terms produces:

$$\frac{C_e}{C_0} = \exp\left[-\frac{0.7 K_T (A_V)^{1.75} L W d}{Q}\right] \tag{6.6}$$

The term LWd/Q represents the hydraulic residence time in an unrestricted flow system. In a free-water-surface wetland, a portion of the available volume will be occupied by the vegetation, so the actual detention time will be a function of the remaining cross-sectional area available for flow. This can be defined as the porosity n of the system in a manner similar to that used for soil systems:

$$n = \frac{V_V}{V}$$

where V_V and V are volume of voids and total volume, respectively.

The product nd is, in effect, the equivalent depth of flow in the system with no vegetation present. In a vegetated system with perfect plug flow, with no extraneous water losses or gains it would also be the ratio of actual residence time to the theoretical detention time. It is also necessary to include an additional factor (the A factor from Eq. 2.1) to account for the settleable BOD, which will be removed near the head of the system in OF and wetland units. Combining these relationships with the general model produces Eq. 6.7.

$$\frac{C_e}{C_0} = A \exp\left[-\frac{0.7K_T(A_V)^{1.75}LWdn}{Q}\right] \tag{6.7}$$

where C_e = effluent BOD, mg/L
C_0 = influent BOD, mg/L
A = fraction of BOD not removed as settleable solids near headworks of the system (as a decimal fraction)
K_T = temperature-dependent rate constant, days^{-1}
A_V = specific surface area for microbial activity, m^2/m^3 (ft^2/ft^3)
L = length of system (parallel to flow path), m (ft)
W = width of system, m (ft)
d = design depth of system, m (ft)
n = porosity of system (as a decimal fraction)
Q = average hydraulic loading on the system, m^3/day (ft^3/day)

Tchobanoglous has suggested that the temperature correction factor for attached growth wetland systems is 1.1.[19] The rate constant K_T (in days^{-1}) at water temperature T (in °C) can therefore be defined by Eq. 6.8

$$K_T = K_{20}(1.1)^{(T-20)} \tag{6.8}$$

where K_{20} is the rate constant at 20°C.

This K_{20} rate constant was determined by applying Eq. 6.6 to the OF systems described by the previously validated Eq. 6.2. The average depth of flow in the system originally studied was 0.5 cm or less, and since the entire surface area is considered "active," the specific surface

A_V can be calculated as 267 m^2/m^3 and the "porosity" of this system as 0.75. This produces a value of 0.0057 day^{-1} for K_{20}, so the general model becomes:

$$\frac{C_e}{C_0} = 0.52 \exp\left[-\frac{0.7K_T(A_V)^{1.75}LWdn}{Q}\right] \tag{6.9}$$

where K_T, the rate constant (in $days^{-1}$) at water temperature T (in °C) is given by

$$K_T = 0.0057(1.1)^{(T-20)}$$

When the bed slope or hydraulic gradient is equal to 1 percent or greater, it is necessary to adjust the design model accordingly:

$$\frac{C_e}{C_0} = 0.52 \exp\left[-\frac{0.7K_T(A_V)^{1.75}LWdn}{4.63S^{1/3}Q}\right] \tag{6.10}$$

where S is bed slope, or hydraulic gradient (as a decimal fraction).

Equations 6.9 and 6.10 look complex, but the expression $LWdn/Q$ can be reduced to the actual detention time t in the system. It is, however, necessary to retain all the components of this expression for design purposes since the dimensions, but not necessarily the detention time for a proposed system, can be defined.

Equations 6.9 and 6.10 fit the OF case very well, but that should be expected since OF data were used to define the rate constant. They were tested with data from the constructed wetland systems at Listowel, Ontario, and Arcata, California,[5,8] and the results are shown in Table 6.3. The volume occupied by the vegetation and the litter was determined to be about 5 percent in the channel wetlands at Listowel. Assuming that the plant stalks are cylindrical and about 1.27 cm (0.5 in) in diameter, it is possible to calculate the specific surface area available for microbial attachment:

$$\text{Surface area of one plant} = 3.14DL$$

$$\text{Volume of one plant} = 3.14D^2L/4$$

Thus in 1 m^3 the vegetation occupies $1 \times 0.05 = 0.05$ m^3.

$$\text{Specific surface} = \frac{\text{surface area}}{\text{volume}} = \frac{4}{0.0127\ \text{m}} \times 0.05 = 15.7\ \text{m}^2/\text{m}^3$$

This value was used in Eq. 6.9 for both the evaluations and is probably reasonable for the general case with a moderately dense stand of emergent vegetation. The porosity n for both the Listowel and Arcata

TABLE 6.3 Predicted versus Actual C_e/C_0 Values for Constructed Wetlands

Location, time period	C_e/C_0 at the distance specified									
	67 m		134 m		200 m		267 m		Final effluent	
	Predicted	Actual	Predicted	Actual	Predicted	Actual	Predicted	Actual	Predicted	Actual
Listowel, Ont.*										
Fall	0.38	0.40	0.27	0.23	0.20	0.19	0.14	0.18	0.10	0.15
Winter	0.40	0.40	0.31	0.20	0.24	0.19	0.18	0.17	0.14	0.17
Spring	0.47	0.30	0.42	0.28	0.38	0.22	0.34	0.23	0.30	0.26
Summer	0.38	0.36	0.27	0.41	0.20	0.30	0.14	0.27	0.10	0.17
Arcata, Calif.†										
Annual average	—	—	—	—	—	—	—	—	0.31	0.28

* Influent BOD = 45 to 56 mg/L; width of cell = 4 m; depth varied with season 11 to 24 cm; Q = 18 to 35 m^3/day; T = 2.8 to 17.8°C.
† Influent BOD = 22 mg/L; width of cell 6.1 m; length = 61 m; depth = 33 cm; Q = 43 m^3/day; T average = 12.2°C.

systems was determined to be 0.75 on the basis of dye study data, and this is probably also a valid assumption for the general case.

The channel wetland unit 4 at Listowel was utilized for the evaluation since it received a higher organic loading, and BOD measurements were made at intermediate points in the 334-m (1095-ft) long wetland during the 1983–1984 test period.[8] Final effluent BODs were the only measurements reported for the Arcata system.[5]

The comparisons in Table 6.3 indicate a reasonably close fit and consistent trends for both the Listowel and Arcata data sets. This would indicate that Eqs. 6.9 and 6.10 can be used for the preliminary design of constructed wetlands with a free water surface as well as for the OF systems described in Chap. 7. Pilot testing to confirm and optimize criteria is recommended for large-scale wetland systems.

Organic loading. The organic loading has been relatively light on most of the constructed wetlands that have been evaluated. The relationship between organic loading and BOD removal rate suggests that a linear correlation exists, at least up to a loading rate of about 100 kg/(ha · day) [89 lb/(acre · day)], which was the highest value reported in the literature for exposed-water-surface wetlands. Most of these wetland systems have been operated at loading rates ranging from 18 to 116 kg BOD/(ha · day) [16 to 104 lb/(acre · day)] and achieve up to 93 percent BOD removal. The organic loading is not used as a principal design criterion for constructed wetlands. It should be considered as a check to ensure maintenance of aerobic conditions in the system, which are important for both wastewater treatment and mosquito control. An upper limit of about 110 kg/(ha · day) [98 lb/[acre · day] is suggested for this purpose.

Since most of the settleable BOD will be removed very close to the inlet point, the organic loading may be very high on this relatively small area. At Listowel, Ontario, for example, the average organic loading over the entire surface area of system 4 was less than 12 kg/(ha · day) [11 lb/(acre · day)], but the organic loading on the first 60-m (197-ft) segment was about 60 kg/(ha · day), with occasional anoxic conditions near the headworks. In the constructed wetland system designed for polishing lagoon effluent in Gustine, California, this problem was solved with wastewater distribution at the head and at the one-third point in the wetland channels and provisions for additional points if needed.[11]

System configuration. Studies in Canada and in California[8,17] confirm that free-water-surface wetland systems should be designed with a large aspect ratio LW to ensure plug-flow conditions and optimum treatment. The Listowel study[8] compared performance of a rectangular wetland ($LW = 4.4:1$) with that of a long narrow channel (L/W 17:1).

The channel received a higher strength wastewater but consistently produced the best effluent. On the basis of these results the Canadian study recommends an aspect ratio of at least 10:1. The 10.5-ha (26-acre) constructed wetland in Gustine, California is divided into 26 parallel cells, each with an aspect ratio of about 20:1.[11]

The depth of water in the system is a critical design parameter to ensure maintenance of aerobic conditions and the design detention time in the system. At a constant hydraulic loading the effect of ET is to increase the detention time beyond the optimum range, resulting in anoxic conditions and effluent deterioration. The work at Listowel[8] suggests that the optimum detention time is somewhere between 7 and 14 days. Effluent quality deteriorated beyond 14 days owing to sluggish flow and anoxic conditions induced by high summer ET losses.

In cold climates the presence of an ice cover on the system will decrease the residence time owing to the reduction in the volume available for wastewater flow. The pilot units in Canada were effectively operated with water depths of about 10 cm (4 in) in the summer and 30 cm (12 in) in the winter. In the latter case the depth was increased in late fall prior to the onset of freezing weather. The expected depth of ice formation should be determined for the local area, and then Eq. 6.9 can be solved for the hydraulic residence time required under the winter water temperature conditions. The depth is then adjusted to provide the necessary residence time (see Chap. 4 for a method of calculating water temperatures). As a rule of thumb it can be assumed that the water temperature will be close to the monthly average air temperature but not less than 2°C (36°F).

The water depth can be increased in all locations with the onset of cool weather. The lower water temperatures should permit higher oxygen levels, and the increased detention time may be needed to compensate for the lower microbial reaction rate. The maximum depth should not exceed 60 cm (24 in) and should preferably be in the range of 30 to 45 cm (12 to 18 in).

Preliminary treatment. Some form of preliminary treatment to at least the primary level is considered necessary for these exposed-water-surface wetland systems to keep the organic loading within reasonable limits and to avoid localized anaerobic conditions. An untested alternative might be to use a grid of elevated sprinklers to apply screened and comminuted wastewater over the first one-third to one-half of the wetland area. This would more evenly distribute the solids and elevate the oxygen content in the water.

The pilot-scale tests in Canada utilized facultative lagoon effluent and unsettled effluent from a 3.5-day detention-time complete-mix aeration cell. Neither proved to be acceptable. The lagoon effluent was high in algae in the summer and in the winter the effluent from the ice-covered lagoon had a very low oxygen content and a high hydrogen sulfide concentration. The high solids level from the aerated cell accumulated at the wetland inlet and caused some dieback of the cattail vegetation.

Algae-laden influent should be avoided since a free-water-surface wetland is not very effective for their removal. The wetland system at Gustine, California is designed to polish facultative lagoon effluent, but the algal problem is managed by selectively using outlets from each of the lagoon cells. In this way the operator can select the lagoon cell with the lowest algal content as the source water for the wetland component.

Conventional primary treatment followed by a brief aeration period to increase oxygen levels would be an acceptable preliminary treatment for a constructed wetland. Imhoff tanks (see Chap. 5 for further discussion) could be used in this manner for small communities. An aeration basin with a 12- to 24-h residence time followed by a settling pond would also be an acceptable alternative. If phosphorus removal is a project objective, the aeration-settling basin combination can serve for this purpose also. If the project must satisfy stringent ammonia limits during the summer months, aeration and recycle of a portion of the effluent are recommended to maintain oxygen levels and sustain the design detention time.

Design criteria summary. For convenience, all the criteria and recommendations developed in the previous sections are summarized below for constructed wetlands with a free water surface.

- Organic loading < 112 kg BOD/(ha · day) [100 lb/(acre · day)]
- Detention time as determined by Eq. 6.9 or 6.10
- Specific surface area (A_V in Eq. 6.9 or 6.10) for attached microbial growth $= 15.7\ m^2/m^3$
- Porosity (n value for Eq. 6.9) of wetland flow path $= 0.75$
- Aspect ratio (L/W) $> 10{:}1$
- Water depth, warm months < 10 cm, cool months < 45 cm

Substituting these factors in Eqs. 6.9 or 6.10 and rearranging terms to solve for detention time or for the required surface area produces the following equations for free-water-surface wetlands:

Hydraulic residence time is given by:

$$t = \frac{(\ln C_0 - \ln C_e) - 0.6539}{65K_T} \tag{6.11}$$

If the bed slope, or hydraulic gradient, is greater than 1 percent, then

$$t = \frac{(\ln C_0 - \ln C_e) - 0.6539}{301 K_T S^{1/3}} \tag{6.12}$$

The design surface area of the wetland is given by

$$A = \frac{Q(\ln C_0 - \ln C_e - 0.6539)}{65 K_T d} \tag{6.13}$$

If the bed slope, or hydraulic gradient, is greater than 1 percent

$$A = \frac{Q(\ln C_0 - \ln C_e - 0.6539)}{301 K_T d} \tag{6.14}$$

where t = hydraulic residence time in the system, days
C_0 = influent BOD concentration, mg/L
C_e = effluent BOD concentration, mg/L
K_T = reaction rate constant, days^{-1}
$= K_{20}(1.1)^{(T-20)}$
K_{20} = 0.0057 days^{-1}
d = design water depth in the system

These equations are only valid for constructed wetlands with a free water surface meeting the conditions defined in the criteria summarized above. Design of OF systems or wetlands with large unvegetated areas must use the general form presented in Eq. 6.9 or 6.10. Example 6.1 below illustrates the design procedure.

Example 6.1 Design a wetland system with a free water surface to produce advanced secondary effluent in a warm climate with a mean annual temperature of 20°C and in a cold climate with summer water temperature at 15°C and winter temperature at 3°C. The design flow is 760 m^3/day (0.2 million gal/day), wastewater BOD is 240 mg/L, and required effluent BOD is 10 mg/L. (*Note:* These are the same conditions used in Example 5.2, so comparisons can be made.)

solution

1. Use an Imhoff tank for primary treatment, followed by a 3-h detention-time aeration basin to increase oxygen levels. The effluent from this preliminary treatment will have a BOD of about 127 mg/L. See Example 5.2 (Chap. 5) for design of Imhoff tanks.
2. Assume the slope of the wetland bed will be 1 percent to allow drainage when required. Use Eq. 6.11 to estimate required detention time.

$$t = \frac{(\ln C_0 - \ln C_e) - 0.6539}{65 K_T}$$

$$\text{At 20°C } t = \frac{(4.8442 - 2.3026) - 0.6539}{(65)(0.0057)}$$

$$= 5.1 \text{ days}$$

At 15°C $t = 8.2$ days

At 5°C $t = 21.3$ days

2. For the warm climate site use a 10-cm water depth on a year-round basis, and use Eq. 6.13 to estimate the surface area required.

$$A = \frac{Q(\ln C_0 - \ln C_e - 0.6539)}{65K_T d}$$

$$\text{At 20°C } A = \frac{(760 \text{ m}^3/\text{day})(4.844 - 2.303 - 0.6539)}{(65)(0.0057 \text{ day}^{-1})(0.10 \text{ m})(10{,}000 \text{ m}^2/\text{ha})}$$

$$= 3.9 \text{ ha}$$

This is about four times the surface area calculated in Example 5.2 for an aerated water hyacinth treatment system. The hyacinth system would require more energy for the continuous aeration and more labor for maintenance and harvesting. The price of land in the area would be a critical factor, and a cost analysis would be necessary for final process selection.

3. Water hyacinths would not be a feasible alternative in the colder climate. Determine the area requirements using Eq. 6.13 and a total water depth of 10 cm in the summer and 30 cm in the winter. Assume ice formation will occupy about 15 cm, so the actual winter depth available for treatment is only 15 cm.

$$\text{At 15°C } A = \frac{(760 \text{ m}^3/\text{day})(4.844 - 2.303 - 0.6539)}{(65)(0.00354 \text{ day}^{-1})(0.10 \text{ m})(10{,}000 \text{ m}^2/\text{ha})} = 6.2 \text{ ha}$$

$$\text{At 3°C } A = \frac{(760 \text{ m}^3/\text{day})(4.844 - 2.303 - 0.6539)}{(65)(0.00113 \text{ day}^{-1})(0.15 \text{ m})(10{,}000 \text{ m}^2/\text{ha})} = 13.0 \text{ ha}$$

The winter conditions control at this site, so the treatment area will have to be 13 ha.

4. Use an aspect ratio of 10:1 and determine the dimensions for the wetland channels for the system in the colder climate, assuming a square plot is available.

$$A = LW = L\frac{L}{10} = \frac{L^2}{10} = 13{,}000 \text{ m}^2$$

Thus $L = 360$ m, $W = 36$ m

Divide the area into 10 parallel channels, each 360 m long. During the summer months operate six of the channels at a water depth of less than 10 cm. Divert enough wastewater to the other channels to maintain plant growth during the summer. Take two channels out of service each summer on a rotational basis for cleaning and routine maintenance.

5. The design procedure assumes and experience at operational systems confirms that SS and nitrogen concentrations in the effluent will also satisfy advanced secondary treatment requirements. Some of the remaining nitrogen will be in the ammonia form. If stringent ammonia limits prevail, recycle of wetland effluent should be incorporated in the system design. The amount of recycle required will depend on the ET losses for the area and should be sufficient to maintain the 8.2-day summer detention time.

Constructed wetlands with subsurface water flow

A constructed wetland with subsurface water flow typically consists of a trench or a bed underlain by impermeable material to prevent seepage and containing a medium that supports the growth of emergent vegetation. The media used have included rock or crushed stone (10 to 15 cm diameter), gravel, and different soils, either alone or in various combinations.

The wastewater flows laterally through the medium and is purified during the contact with the surfaces of the medium and the root zone of the vegetation. This subsurface zone is continuously saturated and therefore is generally anaerobic. However, as discussed in Sec. 6.2, the plants can convey an excess of oxygen to the root system, so there are aerobic microsites adjacent to the roots and rhizomes.

There are several practical advantages for a system with subsurface flow: odors are less likely to occur and mosquitos are less likely to develop. Constructed wetlands with subsurface flow have been used extensively in Switzerland, Denmark, Austria, West Germany, and other European countries and have been described as *root zone method, hydrobotanical system, soil filter trench, biological-macrophytic,* and *marsh beds* by their various proponents. These systems typically use some type of soil as the support medium for the vegetation, and the term *vegetated submerged bed* has been adopted for descriptive purposes in this book. Most of the work in North America has utilized gravel- or rock-filled trenches or beds, and these are described in terms of the media used. The most extensive investigations of this technique in the United States has been conducted as Santee, California with gravel-filled beds.[6]

In the Santee study bulrushes, reeds, and cattails grown in a 76-cm (30-in)-deep gravel bed were compared with each other and with an unvegetated bed for performance with a variety of wastewaters. Typical results using primary effluent (BOD = 118 mg/L, SS = 57 mg/L, NH_3 = 25 mg/L) are given in Table 6.4.

It is clear from the data in Table 6.4 that the deeper root penetration of the bulrush plants made a very significant difference in performance for all the parameters tested. In view of the warm year-round climate at Santee, California and the continuous availability of moisture and

TABLE 6.4 Performance Comparison for Vegetated and Unvegetated Gravel Bed Wetlands at Santee, California[6]

Bed condition*	Root penetration, cm	Effluent quality, mg/L		
		BOD	SS	NH_3
Bulrushes	76	5.3	3.7	1.5
Reeds	> 60	22.3	7.9	5.4
Cattails	30	30.4	5.5	17.7
No vegetation	0	36.4	5.6	22.1

*Q = 3.04 m^3/day, hydraulic residence time = 6 days, bed dimensions: L = 18.5 m, W = 3.5 m, d = 0.76 m.

nutrients for the plants, the depth of penetration observed for cattails and reeds may be close to their maximum practical potential. Further root penetration of the bulrushes was limited by the bed liner, so their maximum potential is not known. These responses suggest that there is little purpose in selecting a design depth for a subsurface-flow wetland that is beyond the potential root penetration depth for the intended emergent vegetation. The aerobic microsites provided by the root zone are essential for effective treatment of applied organics and nitrogen.

Wolverton[21] has conducted bench-scale, batch-type experiments using a wide variety of emergent plants in a rock bed following the equivalent of primary settling. The comparative results are shown in Table 6.5. The influent applied in these experiments was slightly better than primary in quality (after 24 h of settling BOD was 69 mg/L, SS was 28 mg/L, ammonia nitrogen was 11 mg/L, and phosphorus was 6 mg/L).

The use of reeds did provide better performance, but the relationships for the other vegetation is not as clear as the large-scale field study results in Table 6.4. The rate constants that can be derived from this batch laboratory-scale work are considerably higher than comparable values from the Santee project. Systems using these rock beds have been proposed for treating effluent from septic tanks and for polishing lagoon effluent.

A pilot-scale VSB system has been constructed in Ringsted, Denmark to evaluate bed media, bed slope, and vegetation type among other parameters. Typical results during the first 18 months of study for a fine gravel and for a topsoil mixture are shown in Table 6.6. The applied wastewater had been settled but is equivalent in strength to untreated wastewater in the United States.

The results from the Ringsted system do not match the performance

TABLE 6.5 Performance Comparison for Vegetated and Unvegetated Bench-Scale Rock Bed Wetland Units[21]

	Effluent quality, mg/L			
Bed condition*	BOD	SS	NH_3	P
Rushes	7.4	13.9	0.8	4.2
Reeds	2.8	5.9	0.6	2.0
Cattail	10.4	15.3	0.6	3.0
No vegetation	8.2	10.9	8.5	5.5

*Hydraulic residence time = 1 day. Bed dimensions: L = 3 m, W = 0.5 m, d = 0.21 m [16 cm of crushed stone, (2.5–7.5 cm in diameter, overlain by 5 cm of pea gravel, 0.25–1.3 cm in diameter)].

TABLE 6.6 Performance of Pilot-Scale Root Zone Beds at Ringsted, Denmark[9]

	Concentration, mg/L		
		Effluent	
Parameter	Influent	Fine gravel bed*	Soil bed*
BOD	189	11	15
SS	243	6	23
NO_3	0.3	12.6	1.2
NH_3	34	15	15.3
Total N	47.9	29.6	18.8
Total P	15.0	9.6	6.0

*Gravel bed Q = 1.90 m³/day; soil bed Q = 0.821 m³/day, reeds on both beds; bed slope = 8.2% for both; L = 32 m, W = 1.5 m, d = 1 m, gravel = 2–9 mm.

claims of the concept proponents in West Germany.[1] Data from these latter systems are, however, not readily available. The BOD and SS results from Ringsted are comparable with those listed in Table 6.4 for Santee, California.

Hydraulic considerations. The hydraulic regime in these systems is controlled by the permeability or the hydraulic conductivity of the media used and the hydraulic gradient of the system as defined by Darcy's law (Eq. 6.15). In a shallow, fully saturated bed with an impermeable bottom, the hydraulic gradient is for practical purposes the same as the bed slope for grades up to about 6 to 8 percent.

$$Q = k_s A S \tag{6.15}$$

where Q = flow per unit time
k_s = hydraulic conductivity of a unit area of the medium perpendicular to the flow direction
A = cross-sectional area
S = hydraulic gradient of the flow system, $\Delta h / \Delta L$

Consistent units must be used for Q, k_s, and A (see Sec. 3.1 for further discussion of these relationships). Equation 6.15 can be rearranged to solve for the saturated cross-sectional area of the system.

$$A_c = \frac{Q}{k_s S} \tag{6.16}$$

where A_c = dW cross-sectional area of wetland bed, perpendicular to the direction of flow, m² (ft²)
d = bed depth, m (ft)
W = bed width, m (ft)

Q = average flow in the system, m^3/day (ft^3/day)
k_s = hydraulic conductivity of the medium, $m^3/(m^2 \cdot day)[ft^3/(ft^2 \cdot day)]$
S = slope of the bed, or hydraulic gradient (as a decimal fraction)

When a flat bed is used and the gradient is controlled with an overflow weir, use 0.001 or less for S.

On the basis of experience in Europe[2] it is suggested that the unit flow velocity Q/A_c through a cross section of the medium not exceed 8.6 m/day to avoid disruption of the medium-rhizome structure and to ensure sufficient contact time for treatment. This requirement affects the selection of the bed slope S for the system, as shown by Eq. 6.17.

$$S = \frac{8.6}{k_s} \tag{6.17}$$

where k_s is the hydraulic conductivity of the medium.

The design depth of the bed should be selected in accordance with the type of vegetation intended for the system. At present field experience is limited to bulrushes, reeds, and cattails. On the basis of the experience at the Santee, California system the relationships given in Table 6.7 are recommended.

The 60 cm depth for reeds has also been recommended for VSB systems in West Germany,[1] so climate is apparently not a factor in this relationship. The three species listed are common to most temperate zones of the world and can survive almost anywhere. If one particular species seems to dominate the natural wetlands in the general project area, that could then be the basis for selection. Based on experience in Europe it is suggested that the design depths given in Table 6.7 be reduced by 50 percent for industrial wastes with high concentrations of refractory organics (COD).[2]

In one of the VSB systems in use in West Germany[1] it is claimed that about 3 years is necessary for the full 60 cm penetration of the reed rhizomes. They require a partial draining of the bed for a few months in early fall for the first 2 years to induce downward penetration of the rootstock into the moist soil. It is claimed that deeper penetration may not occur if the bed is continuously saturated. A fully

TABLE 6.7 Recommended Design Depth for VSB Wetlands

Plant species	Submerged bed depth, cm
Bulrushes	76
Reeds	60
Cattails	30

developed reed bed with root penetration to 60 cm is claimed to have produced a hydraulic conductivity for the matrix of soil and roots of up to about 260 $m^3/(m^2 \cdot day)$ [853 $ft^3/(ft^2 \cdot day)$] when the original soil was less permeable clayey loam. Such soils have the potential for significant phosphorus removal, but there are insufficient data to verify the claimed hydraulic transformation by root and rhizome development.

It is therefore recommended that the actual horizontal hydraulic conductivity of the soil in place, and not some assumed transformed value, be used for design purposes. In effect this would limit the practical application of this concept to the use of permeable sands and gravels until more performance data become available.

Root penetration is important for the oxygen transfer and treatment of organics and nitrogen at the full bed depth. The maximum treatment potential and possibly the maximum design hydraulic loading may not be realized until the roots and rhizomes have penetrated to their full potential depth.

When gravel or coarse sand is used as the bed medium, rhizome penetration is not necessary to develop the hydraulic properties of the bed. At the reed system in Santee, California the plant roots penetrated to the full 60 cm depth without manipulation of the subsurface water level.

Performance of a gravel-bed system can be expected to improve over the first few years of operation as the root penetration brings oxygen to all parts of the bed. Gravel-bed systems will not, however, be effective for phosphorus removal. The best choice may be the use of moderate- to coarse-textured sands, which should provide some phosphorus removal in most cases.

Once the depth of the bed has been selected and the slope determined (and checked with Eq. 6.17), the bed width can be calculated by Eq. 6.16. This ensures that the design flow can be contained within the bed profile and not emerge as surface flow somewhere down gradient of the application point. Most soil systems have been designed with a slope of 1 percent or slightly higher, but up to 8 percent is acceptable if local terrain permits. The bed slope is also critical for gravel-bed systems to control flow velocity, but the hydraulic gradient in a flat bed can be controlled by the elevation of the inlet and outlet works. In these cases design flow conditions will still be maintained even if the effective hydraulic gradient is 0.10 percent or less.

As shown by Eq. 6.16 the cross-sectional area of the bed is independent of the biochemical reactions in the system and is controlled only by the hydraulic requirements. The length of the bed is the final dimension required; it determines the hydraulic residence time, and for that it is necessary to develop an appropriate design model.

Biological relationships. The removal of BOD in these submerged-bed wetlands can be described with first-order plug-flow kinetics, as described by Eq. 6.1, which is repeated below for convenience.

$$\frac{C_e}{C_0} = \exp\left[-K_T t\right]$$

The hydraulic residence time in the system is a function of the available void spaces V_v and the average flow rate through the system.

$$t = \frac{V_v}{Q} \tag{6.18}$$

where t = hydraulic residence time in the system, days
V_v = volume of voids in the system, m^3 (ft^3)
= nV
= $nLWd$
n = porosity of the bed, as a decimal fraction (see Fig. 2.4 for typical values)
V = total volume of system, m^3 (ft^3)
L = length of system (parallel to flow direction), m (ft)
W = width of system (perpendicular to flow direction), m (ft)
d = depth of submergence, m (ft)
Q = average flow through the system, m^3/day (ft^3/day)

The average flow through the system is equal only to the design hydraulic loading when there are no water losses due to ET or seepage or water gains due to precipitation.

Since the product LW is equal to the surface area of the system, the basic equation can be written as:

$$\frac{C_e}{C_0} = \exp\left(-\frac{K_T A_s d n}{Q}\right) \tag{6.19}$$

where C_e = effluent BOD, mg/L
C_0 = influent BOD, mg/L
K_T = first-order temperature-dependent rate constant, days^{-1}
A_s = surface area of the system, m^2 (ft^2)
n = bed porosity (as a decimal fraction)
d = depth of submergence, m (ft)
Q = average flow through system, m^3/day (ft^3/day)

The temperature dependence of the rate constant is described by Eq. 6.20.

$$K_T = K_{20}(1.1)^{(T-20)} \tag{6.20}$$

where K_{20} = rate constant at 20°C
T = operational temperature of system, °C
K_T = first-order, temperature-dependent rate constant, days^{-1}

Based on limited information from the systems described in Refs. 2, 6, 8, and 9, it appears that the rate constant K_{20} for a particular system may be related to the porosity of the medium used to construct the bed. This is reasonable since natural soils with high porosities tend to have finer void spaces, thereby providing more opportunity for surface contact as compared with a more permeable gravel. The relationship can be tentatively described by Eq. 6.21.

$$K_{20} = K_0(37.31n^{4.172}) \tag{6.21}$$

where K_{20} = the design rate constant at 20°C for the selected bed medium, days^{-1}

K_0 = the "optimum" rate constant for a medium with a fully developed root zone, days^{-1}

= 1.839 days^{-1} for typical municipal wastewaters

= 0.198 days^{-1} for industrial wastewaters with high COD

n = total porosity of the medium selected for bed construction, as a decimal fraction; see Fig. 2.4 for typical values

The root zone VSB systems in West Germany claim a porosity of about 0.42 and a permeability of 260 $m^3/(m^2 \cdot day)$ [850 $ft^3/(ft^2 \cdot day)$] for a bed with full penetration (to 60 cm) of the reed roots and rhizomes.[1] However, minimal performance data are available from these systems to validate the use of Eq. 6.19, 6.20, and 6.21 for finer-textured soils ($n > 0.4$). These equations do adequately describe performance of systems with coarse sands and gravels.

The expression $37.31n^{4.172}$ in Eq. 6.21 is analogous to $C'(A_V)^m$ in Eq. 6.5. The A_V (specific surface area) in the latter expression is an indication of the area in a free-water-surface wetland that is available for aerobic microbial growth. The specific surface area can be estimated for free-water-surface wetlands but is more difficult to determine for the VSB types. In the latter case the available surface area for aerobic microbial growth is indirectly expressed by using the porosity of the bed medium.

An evaluation of Eq. 6.21 with the values from Fig. 2.4 indicates that gravel bed systems will have a kinetic rate constant one-third to one-fourth of that possible for the finer-textured sands. This difference in the kinetic rate constants translates into a much larger surface area requirement for the gravel-bed types, and the economics involved may preclude their adoption in all but relatively small systems. If phosphorus is not a major project concern, perhaps the best choice would be coarse to medium sands with porosities close to 40 percent and permeabilities that would permit full hydrualic loading at project start-up.

Further study is necessary for a complete understanding of the biological kinetics in these VSB wetlands. The relationship described by

Eq. 6.21 and the K_0 values given above are tentative and should only be used for preliminary design estimates. Pilot studies are recommended for final design.

It is possible to determine the surface area required for a VSB wetland by taking the natural logarithm of both sides of Eq. 6.19 and rearranging.

$$A_s = \frac{Q(\ln C_0 - \ln C_e)}{K_T dn} \tag{6.22}$$

where A_s = required surface area of VSB wetland, m^2 (ft^2)
Q = average daily flow through the system, m^3/day (ft^3/day)
C_e = influent BOD, mg/L
C_0 = required effluent BOD, mg/L
K_T = temperature- and porosity-dependent rate constant (determined with Eqs. 6.20 and 6.21)
d = submerged depth of the system, m (ft)
n = porosity of the bed system, as a decimal fraction (see Fig. 2.4 for typical values)

Organic loading. As in the case of free-water-surface wetlands, the surface organic loading is not used as a critical design parameter but rather as a check to ensure that sufficient oxygen will be present in the subsurface bed to sustain the intended treatment reactions. As indicated in Sec. 6.1, the commonly used emergent plants can transmit from 5 to 45 g of oxygen per day per square meter of wetland surface, depending on the oxygen status in the root zone. It is believed that rushes, reeds, and cattails all have roughly the same capability for oxygen transmission; the differing performance noted in various studies is probably due to improper bed design (too deep for the plant selected) or to very high-strength wastewater, which overwhelms the oxygen transfer capabilities of the plant.

The oxygen transfer design for the partial-mix aerated ponds described in Chap. 4 is based on the assumption that the oxygen required is equal to 1.5 times the organic loading (see Example 4.3 and related discussion). It is reasonable to adopt this value for VSB wetlands as well, and to assume a conservative oxygen production rate for the plants of not more than 20 g/(m^2 · day)[0.0061 lb/(ft^2 · day)]. Using these values it is possible to check the oxygen balance in the VSB wetland with Eqs. 6.23 and 6.24.

$$\text{Required } O_2 = 1.5L_0 \tag{6.23}$$

$$\text{Available } O_2 = \frac{(\text{TrO}_2)(A_s)}{1000 \text{ g/kg}} \tag{6.24}$$

where O_2 = oxygen required or available, kg/day (lb/day)

L_0 = organic (BOD) loading, kg/day (lb/day)

TrO_2 = oxygen transfer rate for the vegetation

= 20 g/(m^2 · day) [0.0061 lb/(ft^2 · day)]

A_s = surface area (from Eq. 6.21), m^2 (ft^2)

As a further safety factor it is suggested that the available oxygen as determined from Eq. 6.24 should exceed the oxygen requirements by a factor of 2. In some units processing high-strength wastewaters it may be necessary to increase the operational surface area to achieve this balance. The recommendation that the design bed depths in Table 6.7 be reduced by 50 percent for high-strength industrial wastes is based on the oxygen needs of the system.

At the 20 g/(m^2 · day) oxygen supply rate specified above, the organic loading could be up to 133 kg BOD/(ha · day) [118 lb/(acre · day)], which is in the same range as the value suggested for free-water-surface wetlands. At the theoretical maximum oxygen transfer rate (45 g/m^2 · day) the organic loading could be as high as 300 kg/(ha · day) [268 lb/(acre · day)] on a 60-cm (2-ft) deep bed.

System configuration. A large aspect ratio L/W is not necessary for the VSB wetlands since plug flow will be assured once the wastewater is in the treatment bed (as long as the hydraulic loading does not exceed design limits and allow above-surface flow to occur). In the general case, if sand is used for the bed medium, the width of the bed as calculated with Eq. 6.16 will exceed the length. If finer-textured soil is used, the bed width may be excessive and the resulting bed length unreasonably short. In all cases it is recommended that the bed length (parallel to the flow direction) be at least 20 m (66 ft) to ensure adequate detention time if brief surface overflow does occur.

A critical requirement is to ensure uniform distribution of the wastewater in the entry zone to the treatment bed. This is typically accomplished with a trench filled with crushed rock constructed along the entire width of the bed (see Sec. 6.4 for details).

The design water level is at or just below the treatment bed surface and can be controlled at the outlet works. The bed is designed for average flow conditions, but the typical diurnal variations common to municipal wastewater flow should not be a problem. They may cause some minor surface flow near the inlet for brief periods, but this should quickly infiltrate. If extreme flow variations are expected from industrial operations, it would be prudent to include some equalization capacity in the project design. In the situation in which a community utilizes combined sewers, the peak flow may be very high for brief periods due to the storm water contribution, and surface flow will most certainly occur. It is claimed on the basis of experience with the concept in West Germany that these peak flows could be as high as six times the average flow in dry weather with no significant loss of performance.[2] The vegetation is not harvested in the VSB wetland, so any overland flow will be through the accumulated surface litter, which should provide effective treatment for these brief periods of surface flow.

Preliminary treatment. The VSB wetlands tested in the United States all utilize at least the equivalent of primary treatment as the preliminary treatment level. This might be obtained with septic tanks, Imhoff tanks, conventional primary treatment, or similar systems. A preliminary anaerobic reactor would be useful to reduce the organic and solids content of high-strength industrial wastewaters. If phosphorus removal is a project requirement and gravel or coarse sand has been selected as the treatment bed medium, a preliminary or a post-treatment method for phosphorus removal must be included.

Many of the VSB systems in Europe apply screened and degritted wastewater to the inlet zone of the treatment bed. Solids removal and decomposition are claimed to be very rapid, so clogging does not occur in the inlet zone or in the treatment bed.[1] Any solids accumulations in the inlet trough are periodically removed and also applied to the inlet zone for decomposition. Odors due to the sludge deposits will probably be present in the inlet area if anything of less than primary effluent quality is applied. Odors can be controlled if subsurface distribution to the inlet trench is utilized, but this would require at least primary treatment to avoid clogging.

Example 6.2 Design a wetland system with a submerged treatment bed (VSB) to produce advanced secondary effluent in a cold climate with summer water temperatures at 15°C and winter temperatures at 3°C. Design flow is 760 m^3/day (0.2 million gal/day), wastewater BOD 240 mg/L, and required effluent BOD 10 mg/L. These are the same conditions specified for Examples 5.2 and 6.1 so results may be compared. Cattails are the dominant species in the area and so will be selected as the vegetation for the system.

solution

1. Use an Imhoff tank for primary treatment (see Example 5.2 for design). The effluent from this preliminary treatment will have a BOD of about 127 mg/L.

2. The natural soils can be compacted to provide the necessary impermeable barrier. A medium- to fine-textured sand is locally available and will be used for the bed medium. Porosity is 0.4, and the lateral permeability will be about 350 m/day (this can be measured in a preliminary pilot test and confirmed after

placement of the soil). Design the bed slope at 1 percent for minimal construction effort. Check the suitability of a 1 percent slope with Eq. 6.17.

$$k_s S < 8.60$$

For this case: $(350)(0.01) = 3.5 < 8.60$ (OK)

Since cattails are the selected vegetation, use 30 cm as the design depth of the treatment bed.

3. Determine the cross-sectional area of the bed with Eq. 6.16.

$$A_c = \frac{Q}{k_s S}$$

$$A_c = \frac{760}{(350)(0.01)} = 217.1 \text{ m}^2$$

$$W = A_c/d = 217.1/0.3 = 723.7 \text{ m}$$

These will be the cross-sectional bed dimensions regardless of climate or organic loading since they are controlled by the hydraulic characteristics of the medium.

4. Determine the K_{20} rate constant with Eq. 6.21.

$$K_{20} = K_0\, 37.31 n^{4.172}$$

$$K_0 = 1.839 \text{ days}^{-1} \text{ (since wastewater is moderate-strength municipal type)}$$

$$K_{20} = (1.839)(37.31)(0.40)^{4.172}$$

$$= 1.500 \text{ days}^{-1}$$

5. Determine K_T with Eq. 6.20 for the summer and winter periods.

$$K_T = K_{20}(1.1)^{(T-20)}$$

$$\text{Summer: } K_T = (1.500)(1.1)^{(15-20)}$$

$$= 0.931 \text{ day}^{-1}$$

$$\text{Winter: } K_T = 0.297 \text{ day}^{-1}$$

6. Determine the surface area required with Eq. 6.22.

$$A_s = \frac{(Q)(\ln C_0 - \ln C_e)}{K_T\, dn}$$

$$\text{Summer: } A_s = \frac{(760)(4.84 - 3.00)}{(0.931)(0.30)(0.40)}$$

$$= 12{,}517 \text{ m}^2 = 1.25 \text{ ha}$$

$$\text{Winter: } = 39{,}237 \text{ m}^2 = 3.92 \text{ ha}$$

Winter conditions control, so the total bed area must be 3.92 ha.

7. Determine the bed length L and the detention time t in the system.

$$L = \frac{A_s}{W} = \frac{39{,}237}{724} = 54.2 \text{ m}$$

$$t = \frac{V_v}{Q} = \frac{LWdn}{Q}$$

$$= \frac{(54.2)(724)(0.3)(0.4)}{760} = 6.2 \text{ days}$$

Divide the required area into individual cells 60 m wide for better hydraulic control at the inlet zone. Construct 12 cells, each 60 × 55 m.

All 12 cells must be used in the winter. During the warm months the surface area required is equivalent to only four cells. However, loading only four cells in the summer risks hydraulic overload and surface flow on the cells and desiccation of the "dormant" cells. It is recommended that all 12 remain in service on a continuous basis, but it is clear that there is a significant reserve capacity, and several cells could be shut down during the summer for maintenance if required.

The surface area required for this example is about 25 percent of that required for the free-water-surface wetland in Example 6.1. Had reeds (and a 0.6-m bed depth) been used, the area requirements would have been about 2 ha instead of the 3.9 ha calculated above. A gravel-bed system with reeds might require more than 12 ha, which is about the same area as that required for a free-water-surface wetland.

6.4 Construction Requirements

The unique construction requirements for both types of wetlands include a subsurface flow barrier, selection and placement of the bed media, establishment of vegetation, and the inlet and outlet works. There are several projects in which free-water-surface wetlands have been installed to upgrade wastewater treatment ponds. In these cases one or more of the pond cells are subdivided and converted into channel-type wetlands, with the existing pond bottom serving as the impermeable barrier. This approach would also be feasible for VSB wetlands. To ensure performance reliability and access for maintenance and repair when required, all wetland systems should be subdivided into at least three separate parallel cells (in Example 6.2, the system was divided into 12 units for better hydraulic control at the inlet).

Bed media and liners

Both types of constructed wetlands require an impermeable barrier to ensure containment of wastewater within the system for treatment and to prevent contamination of groundwater. In some cases this may be provided if clay is naturally present or if in situ soils can be compacted to a nearly impermeable state. Chemical treatments, a bentonite layer, asphalt, or membrane liners are also possibilities. This barrier layer must be below the maximum depth of root development for the selected vegetation (see Table 6.7) to avoid penetration of the barrier and leakage. Membrane liners were successfully used at the Santee, California system.[6] A smooth-surfaced, 2-mm membrane (Butyl rubber or plastic) is recommended. Reinforced membranes with a coarse surface or small protuberances may hold in place a rhizome tip, which may then stretch the material to the breaking point as it grows.

The existing topsoil is typically removed prior to construction or installation of the impermeable barrier. If the topsoil has acceptable hydraulic characteristics, it can be reserved and utilized as the bed medium for VSB systems. If phosphorus removal is not a project requirement, the VSB bed material could be medium- to coarse-textured sand. The type of soil is not critical for a free-water-surface wetland, but a soil depth above the impermeable barrier that is at least equal to the maximum root penetration is necessary. The depth of material placed for a VSB wetland should be equal to the design depth plus another 10 to 15 percent to allow for any consolidation that might occur.

The bed material should be placed on top of the impermeable barrier with light equipment to avoid puncture of the barrier or excessive compaction of the material. After the material has been spread and brought to grade, a light scarification or cultivation to break up any compacted zones is recommended for the VSB types.

Vegetation

As shown in Table 6.7, it is possible to utilize a greater bed depth for a VSB system if reeds or bulrushes are selected as the vegetation type. However, to ensure successful establishment and maintenance over time it is suggested that the species that dominates in the local area be selected for use and the design depth adjusted accordingly. Vegetation selection for the free-water-surface wetlands should also be based on local conditions.

The vegetation will propagate from seed, so aerial seeding might be considered for large-scale systems. Plant development from seed takes a significant time, and project start-up may be delayed. The quickest and most reliable approach is to transplant rhizomes of the vegetation of choice in the prepared treatment bed. Rhizomes can be obtained from some commercial nurseries or harvested directly from natural wetlands in the local area. At the wetland in Listowel, Ontario cattail rhizomes planted on 1 m (3 ft) centers produced a dense stand within 3 months. About the same spacing should also be suitable for bulrushes. The reed plant has a thinner leaf canopy, so a closer spacing is necessary [50 cm (20 in) centers].

Each rhizome cutting (about 10 cm long), which should have at least one bud, is planted with one end about 4 cm (2 in) below the surface of the medium and the other end exposed to the atmosphere. Planting can take place after the last frost in the spring or before the first frost in the fall. The bed is then flooded and the water level maintained just below the surface of the medium for the next several months until significant new growth has developed and emerged. At this stage the free-water-surface wetlands and the VSB systems with sand or coarser material can be put into full-scale operation.

In the VSB systems with finer-textured soils it may not be possible to apply the full design hydraulic loading (without surface flow) until the plant rhizomes have completely penetrated the bed. This may take up to 3 years and will require manipulation of the water level in the bed to encourage deeper penetration of the rhizomes. Typically, the water level is lowered to about one-third of the bed depth in September and held at that level for 2 to 3 months. Assuming a spring planting, this water level adjustment would take place during the first and second fall seasons following system construction. This procedure takes advantage of the normal growth cycle of the plant. In the spring and summer most of the growth is directed toward the aboveground emergent portions. In the fall, growth of the rhizomes occurs, and their vertical penetration is encouraged by moist rather than saturated soil conditions.

Inlet and outlet structures

Inlet and outlet structure requirements are significantly different for the two types of constructed wetlands discussed in this chapter. In free-water-surface wetlands proper distribution of organics and the potential for mosquito problems must be considered. The VSB types must provide uniform hydraulic loading and effluent collection to ensure utilization of the full treatment bed.

The free-water-surface wetlands require the same concerns with respect to organic loading and mosquitos as the hyacinth systems discussed in Chap. 5 (see Figs. 5.2 and 5.3 and the related discussion on structural elements and mosquito control for recommendations that are applicable to this type of wetland also). The free-water-surface wetlands should be constructed on at least a 1 percent grade so that the system can be drained. The outlet could be a stop-log weir for this purpose, or a sump could be constructed near the outlet point for pumping the final increment of water.

The critical requirement for VSB wetlands is to ensure the uniform distribution of the wastewater in the entry zone to the treatment bed. This is typically accomplished with a trench filled with crushed rock constructed along the entire width of the bed. This entry-zone trench might be up to 3 m (10 ft) wide and equal in depth to the treatment bed. The crushed rock should be in the same size range as that used for trickling filters, 6 to 10 cm (2 to 4 in) in diameter. Wastewater can be applied to the entry zone from a parallel trough with typical V-notch overflow weirs or by similar methods that will ensure uniform distribution. This entry zone is also planted with the same vegetation used on the treatment bed. The weir crest should be about 0.5 m (1.5 ft) above the bed surface to allow for the accumulation of sludge and plant detritus.

Effluent collection is also important to ensure full utilization of the entire bed area. An outlet trench, similar to that described above, is placed across the entire width at the end of the treatment bed. The trench would typically contain a perforated pipe for effluent collection. This pipe should connect to some sort of adjustable outlet so that the water level in the bed can be controlled.

6.5 System Management

The special requirements for constructed wetlands include management of the vegetation and other solid residues and control of mosquitos. The routine operation and maintenance procedures for the free-water-surface wetlands are similar to those for the pond systems described in Chap. 4. Muskrats and other burrowing rodents can be a problem for earthen dikes and berms.

Vegetation

The vegetation is a critical component in both free-water-surface and VSB wetlands. A relatively dense stand of healthy plants is essential for the successful performance of these treatment systems. Once the vegetation is established, the continuous supply of moisture and nutrients provided by the wastewater should ensure successful growth and reproduction for many years. The plants are susceptible to damage from high concentrations of toxic wastes, and dieback will occur if the oxygen demand in the soil significantly exceeds the oxygen transmission capacity of the plant for a significant period of time. If major plant losses occur, the area must be revegetated to restore the intended treatment capability.

In the fall of each year the emergent portion of the plant will die, and this is the major growth period for the buried rhizomes and roots. An issue of concern is whether the emergent portion of the plant needs to be harvested and at what frequency. If the plant is not harvested, it will decompose and release organics and nutrients. In natural wetlands this usually results in peak levels of nitrogen and phosphorus in the spring; in constructed wetlands this release is typically a small fraction of the organic and nutrient loading on the system and can be ignored. Since, as described in Sec. 6.1, the plants remove a minor portion of the total nutrients removed by the system, their harvest for this purpose is not required.

An annual harvest is necessary for the reed bed systems described in Chap. 8, which are utilized for sludge dewatering. This is necessary so that the plant litter does not interfere with proper distribution of the sludge when it is applied to the bed. The same principle should also hold for wastewater treatment in free-water-surface wetlands. In this case harvesting only needs to be considered when the accumulation of litter interferes with the hydraulic regime in the bed and causes short circuiting of flow or "dead spots." The frequency of such harvests might range from every 7 to 9 years in warm climates to 2 to 5 years in colder climates, where the decomposition rates for the plant litter are slower.

The capability for access and harvest should be included in all free-water-surface wetland designs. This need for periodic harvest and removal of accumulated surface litter is one of the reasons the treatment system should be divided into multiple parallel cells. One or more cells can be taken out of service in the warm months, allowed to dry, and then cleaned. Controlled burning can be an alternative to plant harvest for these free-water-surface wetlands and can be utilized without damaging the subsurface portions of the plant.

Routine harvests are not necessary and should not be included in the management plan for the VSB wetland types, as the accumulation of plant litter above the ground surface will not interfere with the subsurface flow in these systems. This accumulated litter provides a significant degree of thermal insulation, thereby permitting continued winter operation of the system in cold climates. Since the VSB systems are totally dependent on oxygen transmission by the plants to the root zone, it is important not to disrupt this activity with harvests or other manipulations.

Plant litter and sludge will accumulate on and near the inlet zone for these VSB wetlands and their eventual removal is required, so access for this purpose must be provided. The frequency will depend on the elevation difference between the bed surface and the inlet weir; it might be 20 years or more for a weir height of 0.5 m (1.6 ft).

Mosquito control

Mosquitos are not a concern for the VSB wetlands as long as the subsurface flow regime is maintained. When surface overflow occurs, mosquitos may be a minor problem for brief periods if any residual ponding occurs.

Mosquito control is an essential element in all but the most remote free-water-surface wetlands. The same procedures described in Chap. 5 for hyacinth systems are also applicable to this type of wetland. *Gambusia* fish provide effective control in climates in which winter freezing is not a factor. The fish require aerobic conditions and will not enter anoxic zones. This is one of the reasons for the special attention given to the inlet and outlet works, as described in Chap. 5 and in Sec. 6.4 above. Recirculation of a portion of the wetland effluent during the warmest summer months can also be used to maintain desirable oxygen levels and flow conditions.

Removal of accumulated plant litter also contributes to mosquito control by eliminating dead spots and allowing access for the *Gambusia*. Temporarily draining a treatment cell can also be used to interrupt the breeding and development cycle of the insect. The pilot study in Arcata, California[5] successfully used both *Gambusia* and a pupaecide (Altosid) for mosquito control. The *Gambusia* would not survive the low temperature and low oxygen levels in an ice-covered wetland in a cold climate, so other measures are required. At the system in Listowel, Ontario[8] the bacterial insecticide *Bacillus thuringiensis* var. *israeliensis* provided effective control and was recommended for use elsewhere.

The side slopes of the containing dikes or berms should be as steep as possible and any vegetation on these surfaces should be controlled. The presence of duckweed (*Lemna* spp.) in the wetland will also contribute to mosquito control, as described in Chap. 5; however, an extensive duckweed mat will also interfere with atmospheric reaeration of the water.

REFERENCES

1. Boon, A. G.: *Report of a Visit by Members and Staff of WRC to Germany to Investigate the Root Zone Method for Treatment of Wastewaters,* Water Research Centre, Stevenage, England, August 1985.
2. Boon, A. G.: *Report of a Visit by A. G. Boon to Canada and the USA to Investigate the Use of Wetlands for the Treatment of Wastewater,* Water Research Centre, Stevenage, England, March 1986.
3. Dinges, R.: *Natural Systems for Water Pollution Control,* Van Nostrand Reinhold, New York, 1982.
4. Eckenfelder, W. W., Jr.: *Water Quality Engineering for Practicing Engineers,* Barnes & Noble, New York, 1970.
5. Gearheart, R. J., S. Wilbur, J. Williams, D. Hull, B. Finney, and S. Sundberg: *Final Report City of Arcata Marsh Pilot Project Effluent Quality Results—System Design and Management,* Project Report C-06-2270, City of Arcata, Department of Public Works, Arcata, Calif., 1983.
6. Gersberg, R. M., B. V. Elkins, S. R. Lyons, and C. R. Goldman: "Role of Aquatic Plants in Wastewater Treatment by Artificial Wetlands," *Water Res.,* 20:363–367, 1985.
7. Godfrey, P. J., E. R. Kaynor, S. Pelczarski, and J. Benforado: *Ecological Considerations in Wetlands Treatment of Municipal Wastewaters,* Van Nostrand Reinhold, New York, 1985.
8. Herskowitz, J.: *Town of Listowel Artificial Marsh Project Final Report,* Project No. 128RR, Ontario Ministry of the Environment, Toronto, September 1986.
9. Jacobsen, B. N.: *Physical Description of the Root Zone Installation in Ringsted and Rodekro Municipalities, Experimental Plan and Preliminary Results,* Water Quality Institute (VKI), Horsholm, Denmark, 1985.

10. Lawson, G. J.: *Cultivating Reeds* (Phragmites australis) *for Root Zone Treatment of Sewage,* Project Report 965, Institute of Terrestrial Ecology, Cumbria, England, October 1985.
11. Nolte & Associates: *Marsh System Pilot Study City of Gustine, California,* Environmental Protection Agency Project No. C-06-2824-010, Nolte & Assoc., Sacramento, Calif., November 1983.
12. Otta, J. W., T. G. Searle, and S. V. Gaddes: "Land Treatment Enhances Habitat of the Endangered Mississippi Sand Hill Crane," *Proc. Third Water Reuse Symp.,* Am. Water Works Assoc., Denver, Colo., 1984, pp. 649–659.
13. Reed, S. C., and R. K. Bastian (eds.): Aquaculture Systems for Wastewater Treatment: An Engineering Assessment, Environmental Protection Agency 430/9-80-007, National Technical Information Service, PB81156689, 1980.
14. Reed, S. C., and R. K. Bastian: "Wetlands for Wastewater Treatment: an Engineering Perspective," *Ecological Considerations in Wetlands Treatment of Municipal Wastewaters,* Van Nostrand Reinhold, New York, 1985, pp. 444–450.
15. Reed, S. C., C. J. Diener, and P. B. Weyrick: *Nitrogen Removal in Cold Regions Trickling Filter Systems,* SR 86-2, Cold Regions Res. & Eng. Lab., Hanover, NH, February 1986.
16. Stephenson, M., G. Turner, P. Pope, J. Colt, A. Knight, and G. Tchobanoglous: *The Use and Potential of Aquatic Species for Wastewater Treatment, Appendix A, The Environmental Requirements of Aquatic Plants,* Publication No. 65, California State Water Resources Control Board, Sacramento, Calif., 1980.
17. Stowell, R., S. Weber, G. Tchobanoglous, B. Wilson, and K. Townzen: "Mosquito Considerations in the Design of Wetland Systems for the Treatment of Wastewater," *Ecological Considerations in Wetlands Treatment of Municipal Wastewaters,* Van Nostrand Reinhold, New York, 1985, pp. 38–47.
18. Tchobanoglous, G.: *Wastewater Engineering Treatment Disposal Reuse,* McGraw-Hill, New York, 1979.
19. Tchobanoglous, G., and G. Culp: *Aquaculture Systems for Wastewater Treatment: An Engineering Assessment,* EPA 430/9-80-007, Environmental Protection Agency Office of Water Program Operations, Washington, D.C., June 1980, pp. 13–42.
20. U.S. Environmental Protection Agency: *Process Design Manual for Land Treatment of Municipal Wastewater,* EPA 625/1-81-013, Center for Environmental Research Information, Cincinnati, Ohio, Oct. 1981.
21. Wolverton, B. C., R. C. McDonald, and W. R. Duffer: "Microorganisms and Higher Plants for Wastewater Treatment," *J. Environ. Qual.,* **12(2)**:236–242, 1983.
22. Zirschky, J.: *Basic Design Rationale For Artificial Wetlands,* Contract Report 68-01-7108, Environmental Protection Agency, Office of Municipal Pollution Control, Washington, D.C., June 1986.

Chapter

7

Land Treatment Systems

Land treatment processes include slow rate (SR), overland flow (OF), and rapid infiltration (RI). In addition to these three processes, land is also used for various on-site soil absorption systems used to treat septic tank effluent. References 43 and 45 should be consulted for these latter methods.

7.1 Systems Types

Land treatment is the controlled application of wastewater to the soil to achieve treatment of constituents in the wastewater. The three types of land treatment systems all use the natural physical, chemical, and biological processes within the soil-plant-water matrix. The SR and RI processes utilize the soil matrix for treatment after infiltration of the wastewater, the major difference between the processes being the rate at which the wastewater is loaded onto the site. The OF process uses the soil surface and vegetation for treatment, with the treated effluent collected as runoff. The characteristics of these systems are compared in Table 7.1, and performance expectations are given in Table 1.3.

Slow rate systems

Slow rate systems are the predominant form of land treatment of municipal and industrial wastewater. The technology is similar to that

TABLE 7.1 Characteristics of Terrestrial Wastewater Treatment Systems

Characteristic	System type		
	Slow rate	Overland flow	Rapid infiltration
Application method	Sprinkler or surface	Sprinkler or surface	Usually surface
Minimum preapplication treatment	Primary	Fine screening	Primary
Annual loading rate, m/year	0.5–6	3–20	6–125
Disposition of applied wastewater	Evapotranspiration and percolation	Surface runoff and evapotranspiration	Percolation

of conventional agricultural irrigation, and the loading rates, as shown in Table 7.1, are the lowest of the land treatment methods. However, as discussed in Chap. 2, SR has the widest range of acceptable soil types and permeabilities. A list of large municipal SR systems is presented in Table 7.2.

Slow rate systems have been traced back to 1531 in Bunzlau, Germany and 1650 in Edinburgh, Scotland.[19] The practice was well organized in England from the 1850s through the 1870s.[13] In the 1880s many U.S. municipalities were reported to be using wastewater for irrigation, a practice that started in 1872 in Augusta, Maine.[28] The large SR system in Melbourne, Australia was established in 1897.[31]

TABLE 7.2 Selected Municipal Slow Rate Land Treatment Systems

Location	Flow, m^3/day	System area, ha	Application method
Bakersfield, Calif.	73,600	2,060	Ridge and furrow and border strip surface application
Clayton Co., Ga.	75,950	960	Solid-set sprinklers
Lubbock, Tex.	62,500	2,000	Center-pivot sprinklers
Mitchell, S.D.	9,300	520	Center-pivot sprinklers
Muskegon Co., Mich.	110,400	2,160	Center-pivot sprinklers
Petaluma, Calif.	20,000	220	Traveling-gun sprinklers
Vernon, B.C.	10,300	591	Traveling-gun and side-roll sprinklers

There are now over 800 operating slow rate systems in the United States.[41]

Overland flow systems

Overland flow is a land treatment process in which wastewater is treated as it flows down carefully graded grass-covered slopes. In contrast to the SR process, in which surface runoff is avoided, in the OF method surface runoff is a design requirement, the treated runoff being collected at the bottom of the slopes. To achieve the required runoff the soils must either be slowly permeable, be compacted during construction to limit percolation, or have an impermeable layer just below the surface. Wastewater is either sprinkler- or surface-applied to the top of the slope, and treatment occurs during the slow travel of the water in thin sheet–flow down the slope. The slopes are typically 2 to 8 percent in grade and 30 to 61 m (100 to 200 ft) in length. The features of an OF system are shown in Fig. 7.1.

The OF process evolved in the United States from "spray runoff" as practiced with food processing wastewater[28] to an advanced treatment process capable of being designed for removal of biochemical oxygen demand (BOD), suspended solids (SS), and nitrogen.[39] Modifications to achieve significant phosphorus removal by precipitation with added

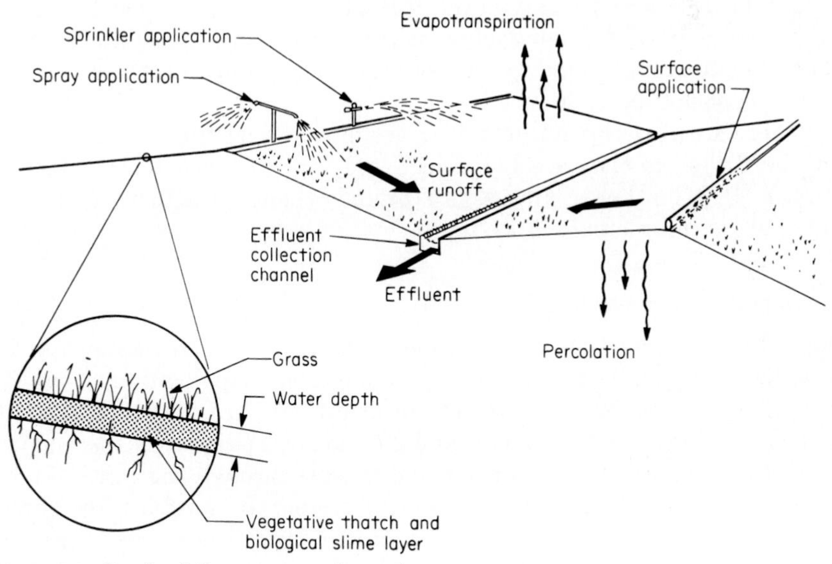

Figure 7.1 Overland flow process schematic.

TABLE 7.3 Municipal and Industrial OF Systems in the United States

Municipal systems	Industrial systems
Alum Creek Lake, Ohio	Chestertown, Maryland
Beltsville, Maryland	Davis, California
Carbondale, Illinois	El Paso, Texas
Cleveland, Mississippi	Middlebury, Indiana
Corsicana, Texas	Napoleon, Ohio
Davis, California	Paris, Texas
Falkner, Mississippi	Rosenberg, Texas
Gretna, Virginia	Sebastopol, California
Heavener, Oklahoma	Woodbury, Georgia
Lamar, Arkansas	
Minden-Gardnerville, Nevada	
Mt. Olive, New Jersey	
Newman, California	
Norwalk, Iowa	
Vinton, Louisiana	

alum on the OF slopes have been researched.[38] A process used in England and Australia known as *grass filtration* is in essence an OF process, and the constructed wetlands described in Chap. 6 are also similar in concept and performance.

There are about 50 municipal OF systems in the United States. Some of the full-scale ones are listed in Table 7.3. The research conducted in the 1970s on the OF process is expected to contribute to a significant increase in municipal OF systems.

Rapid infiltration systems

Rapid infiltration is a land treatment process in which wastewater is treated as it percolates through a permeable soil. Applications, usually to shallow spreading basins, are intermittent. Treatment is accomplished by physical, chemical, and biological processes as the wastewater infiltrates the soil surface and travels through the soil.

Vegetation is usually not a part of RI systems because loading rates are too high for nutrient uptake to be effective. There are, however, situations in which vegetation can play an integral role in stabilizing surface soils and maintaining high infiltration rates.[30]

The treated water from most of the 300 municipal RI systems flows through the subsurface until it joins a surface water body. This indirect surface discharge is generally encouraged by regulatory agencies, as opposed to permanent groundwater discharge to drinking water aquifers. There are, however, installations in which the percolate is recovered by pumping, such as at Phoenix, Arizona and in the Dan region of Israel.[3,18]

7.2 Slow Rate Systems

Design objectives and process design procedures for SR systems are described in this section; details on site selection for these systems are presented in Chap. 2.

Design objectives

There are basically two types of SR systems. Type 1 systems, designed on the basis of the *limiting design factor* (LDF) concept, apply the maximum possible amount of wastewater to the minimum possible land area. The critical parameter or factor that limits the loading rate is identified for the specific site and the particular wastewater by comparing the wastewater loadings allowed for each constituent. For municipal SR systems the LDF is usually the hydraulic capacity of the soil profile or the nitrogen content of the wastewater. For industrial SR systems the LDF may be hydraulic capacity, nitrogen, BOD, metals, or, in the case of toxic wastes, the primary toxic or hazardous constituent.[23]

The type 2 SR system is designed to optimize the water reuse potential. In this case just enough water is applied to satisfy the total irrigation requirements for the crop being grown. The water loading rate sets the land area requirement and depends on the climate, the soil, the crop, the leaching requirements, and the method of irrigation. The basic intent with these systems is to irrigate the maximum possible amount of land.

Preapplication treatment

The treatment needed prior to SR land treatment depends on the type of wastewater, the type of crop, the degree of public access to the site, and the percolate quality requirements. For municipal wastewater the main concern is to reduce the pathogen content of the wastewater and to minimize the nuisance potential by providing at least primary treatment. For industrial wastewater the preapplication treatment varies

TABLE 7.4 Guidance on Preapplication Treatment for Municipal Slow Rate Systems

A. Primary treatment—acceptable for isolated locations with restricted public access and when limited to crops not for direct human consumption.
B. Biological treatment by lagoons or in-plant processes, plus control for fecal coliform count to less than 1000 MPN*/100 mL—acceptable for controlled agricultural irrigation except for human food crops to be eaten raw.
C. Biological treatment by lagoons or in-plant processes, with additional BOD or SS control as needed for aesthetics, plus disinfection to log mean of 200/100 mL (EPA fecal coliform criteria for bathing waters)—acceptable for application in public access areas such as parks and golf courses.

*MPN = most probable number.

with the type of wastewater and may include fine screening, pH adjustment, sedimentation, and/or grease removal. Guidance for assessing preappliction treatment is presented in Table 7.4.

Preapplication treatment for most municipal SR systems consists of biological treatment in ponds. Ponds are generally a cost-effective method of treatment and can also provide some of the storage volume needed in most SR systems. Ponds can provide removal of fecal coliforms, as described in Chap. 3, and can effectively reduce nitrogen concentrations, as described in Chap. 4. The latter is particularly important since nitrogen is often the LDF for municipal SR systems.

Crop selection

The crop is very important in the SR process because it removes nitrogen, maintains or increases wastewater infiltration rates, and can produce revenue, particularly in type 2 systems. In type 1 systems, in which wastewater application rates are maximized, the crop is often selected to maximize nitrogen removal or withstand high hydraulic loading rates. Nutrient uptake rates for forage and field crops are presented in Table 7.5, and nitrogen uptake rates for forest ecosystems are presented in Table 7.6. Nitrogen uptake is a function of crop yield as well as nitrogen content of the harvested portion of the crop. As a result, in climates where the yield of the crop's dry matter is high, the total nitrogen removal (in kilograms per hectare) will be higher than in colder climates with shorter growing seasons.

Legume crops can fix nitrogen from the air; however, they are active scavengers for nitrate nitrogen if it is present in the soil. As a result, if legumes are fertilized with nitrogen, most of the crop uptake will come from the fertilizer or wastewater.

In tropical or subtropical climates forages such as bahia grass or California grass can be grown, which will take up substantially more nitrogen than the forages listed in Table 7.5. For example, California

TABLE 7.5 Nutrient Uptake Rates for Selected Crops[41]

Crop	Nutrient, kg/(ha · year)		
	Nitrogen	Phosphorus	Potassium
	Forage Crops		
Alfalfa*	225–675	22–34	174–224
Bromegrass	130–224	40–56	247
Coastal bermuda grass	400–675	35–45	225
Kentucky bluegrass	200–270	45	200
Quack grass	235–280	30–45	275
Reed canary grass	335–450	40–45	315
Ryegrass	200–280	60–85	270–325
Sweet clover*	175–300	20–40	100–300
Tall fescue	150–325	30	300
Orchard grass	250–350	20–50	225–315
Timothy	150	24	200
Vetch	390	46	270
	Field Crops		
Barley	125–160	15–25	20–120
Corn	175–250	20–40	110–200
Cotton	75–180	15–28	40–100
Grain sorghum	135–250	15–40	70–170
Oats	115	17	120
Potatoes	230	20	245–325
Rice	110	26	125
Soybeans*	250–325	10–28	30–120
Sugar beets	255	26	450
Wheat	160–175	15–30	20–160

* Legumes may also take up a minimal amount of nitrogen from the atmosphere when under nitrogen fertilization.

grass has been found to remove 2000 kg/(ha · year) [1780 lb/(acre · year)] in field experiments.[16]

Other crop characteristics of importance in addition to nitrogen uptake are evapotranspiration (ET), water tolerance, salinity tolerance, and revenue potential. Evapotranspiration is the consumptive use of water (both evaporation and transpiration) by the growing crop. The ET rate is controlled by atmospheric conditions and soil water availability. If sufficient soil water is available, the potential ET will be determined by solar radiation, air temperature, wind speed, and relative humidity. A map of average potential ET and mean annual precipitation is presented in Fig. 7.2.

For forage grasses and trees the actual ET will be nearly the same as the potential ET. For field crops the actual ET will usually be less than the potential ET, especially at the beginning and end of the growing season. Estimates of potential and actual ET values in most western states can be obtained from local agricultural extension offices,

TABLE 7.6 Nitrogen Uptake for Selected Forest Ecosystems[4,41]

Forest type	Tree age, years	Annual nitrogen uptake, kg/(ha · year)
Eastern Forests		
Mixed hardwoods	40–60	220
Red pine	25	110
Old field with white spruce	15	280
Pioneer succession	5–15	280
Southern Forests		
Mixed hardwoods	40–60	340
Southern pine without understory	20	220*
Southern pine with understory	20	320
Lake State Forests		
Mixed hardwoods	50	110
Hybrid poplar†	5	155
Western Forests		
Hybrid poplar†	4–5	300–400
Douglas fir plantation	15–25	150–250
Slash pine	20	370

* Principal southern pine included in these estimates is loblolly pine.
† Short-term rotation with harvesting at 4 to 5 years; represents first growth cycle from planted seedlings.

research stations, or the U.S. Soil Conservation Service (SCS). Potential ET values can be estimated from temperature and other climatic data.[11,29]

Selection of the crop for a type 1 SR system should focus on nitrogen removal, compatibility with hydraulic loadings (tolerance of overirrigation), and ease of management (minimal cultivation and harvesting requirements). Consideration of all these factors usually leads to the selection of forage or tree crops as best suited for type 1 SR systems.

The important crop selection criteria for type 2 systems are water requirements, revenue potential, compatibility with local climate and soils, and salinity tolerance. Field crops are usually chosen because of their revenue potential and compatibility with existing local practices. The tolerance for salinity must be considered for field crops because some are sensitive to total dissolved solids values over 700 mg/L. Tolerance for chlorides and boron needs also to be considered for field and fruit crops.[44]

Loading rates

Most SR systems are limited either by hydraulic or nitrogen loading rates. The hydraulic loading rate for type 1 systems is based on the

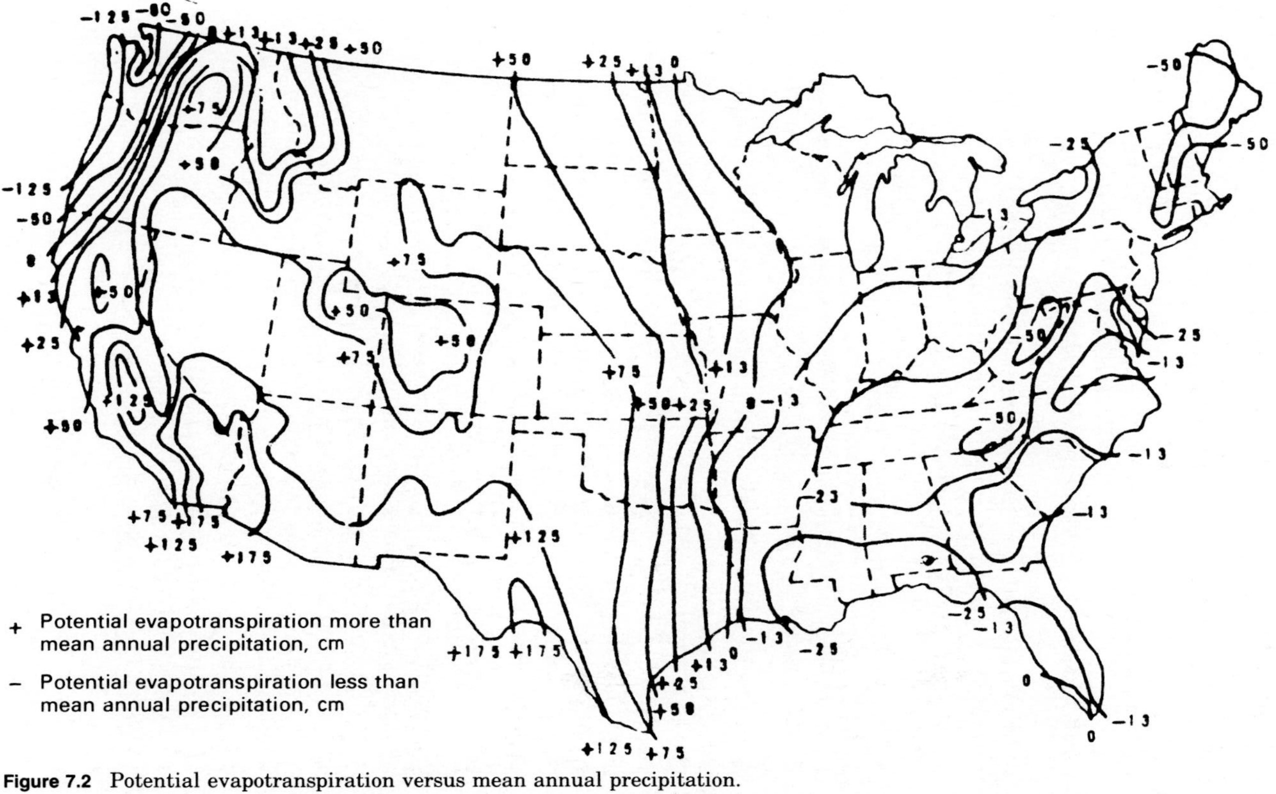

Figure 7.2 Potential evapotranspiration versus mean annual precipitation.

soil permeability. Hydraulic loading rates are expressed in centimeters per week or meters per year (inches per week or feet per year) to reflect an average loading over a hydraulic loading cycle that includes application periods and drying periods.

For type 2 systems the hydraulic loading rate is based on irrigation requirements, which reflect crop ET rates and a leaching (percolation) fraction that is used to prevent buildup of salts in the soil profile. Nitrogen and BOD loading rates as well as any unique constituents should be checked for industrial wastewater systems.

Hydraulic loading for type 1 SR systems. The water balance equation is the basis of the hydraulic rate determination:

$$L_w = \mathrm{ET} - P_r + P_w \tag{7.1}$$

where L_w = wastewater hydraulic loading rate
ET = evapotranspiration rate
P_r = precipitation rate
P_w = percolation rate

(Note that units must be consistent for all terms of the equation.) Any surface runoff is assumed to be captured and reapplied. The water balance is often used on a monthly basis to determine the allowable wastewater loading rate based on the permeability of the limiting layer in the soil profile (See Chap. 3). To determine the monthly balance the design values of precipitation and ET must be determined; the wettest year in 10 is often adopted as the design year for this purpose.

To determine the minimum permeability of the soil profile it is usually necessary to conduct field tests. Acceptable tests include the basin flooding technique[41] and the use of cylinder infiltrometers, sprinkler infiltrometers, or air entry permeameters.[29,41] An average percolation rate can be calculated as described in Sec. 3.1.

The design percolation rate P_w is calculated from the length of the application period in comparison with the overall cycle of wetting and drying and other factors such as the variability of soil conditions. With use of either SCS permeability ranges or field test results it is recommended that the daily design percolation rate should range from 4 to 10 percent of the measured or published rate. If the scheduled frequency of application is one day per week on a given portion of the application site, the 4 to 10 percent value would be incorporated as follows:

$$P_w(\text{daily}) = K(24\ \text{h/day})(0.04\ \text{to}\ 0.10) \tag{7.2}$$

where P_w = design percolation rate, cm/day (in/day)
K = permeability of limiting soil layer, cm/h (in/h)
0.04 to 0.10 = adjustment factor to account for wet/dry ratio and ensure a conservative value for infiltration of wastewater

The adjustment factor depends on the variability of the site soils and the wet/dry ratio. The wet/dry ratio is 0.15 or less for most SR systems and will not affect the adjustment factor unless it is less than 0.04. The lower end of the range (0.04) should be used for the adjustment factor when the soil permeability values are variable and the wet/dry ratio is low, and 0.10 should be used as the adjustment factor when soil permeability values are relatively uniform and the wet/dry ratio is 0.10 or higher.

The design rate calculated by Eq. 7.2 assumes that the wastewater infiltrates at the measured rate K but for only a portion of the month. During months with high rainfall the wet/dry ratio for the site changes, and more total water will infiltrate. Since type 1 systems are designed for maximum wastewater application, the water balance equation is modified, and the monthly wastewater P_w is still applied in the months when precipitation exceeds ET if weather and crop conditions permit. In the general case all the precipitation and wastewater will still infiltrate even in the wet months because of the very conservative adjustment factors used in determination of P_w. Operational adjustments may be necessary to avoid wastewater application during intense rainstorms. Equation 7.1 is applied directly during the months when ET exceeds precipitation, so the monthly wastewater loading can be increased above the P_w value to make up for the ET deficit.

The monthly percolation can then be determined by multiplying the daily value of P_w by the number of application days per month. Because the precipitation must also percolate (or be lost to ET), any downtime for precipitation should not be included in the water balance. However, the downtime for harvesting, planting, or wintertime soil freezing must be included. Example 7.1 demonstrates the procedure for developing a project water balance and the design hydraulic loading for a type 1 SR system.

Example 7.1 Determine the monthly water balance and the design hydraulic loading rate based on soil permeability. Assume that the soil profile has a moderately slow permeability of 0.5 cm/h: the site is in a relatively warm climate but operating days are restricted by freezing temperatures to 10 days in January, 12 days in February, 15 days in March, 15 days in November, and 10 days in December; precipitation (P_r) and ET records are available from local agencies. Assume a grass forage crop so cultivation is not required but harvesting will require 5 days in July and in September.

solution

1. Determine the allowable daily percolation rate for the applied wastewater with an assumed adjustment factor of 0.07.

$$P_w = K(24 \text{ h/day})(0.07)$$

$$P_w = (0.05)(24)(0.07) = 0.84 \text{ cm/day}$$

2. Tabulate ET and precipitation values and determine net loss or gain:

Month	ET, cm/month	P_r, cm/month	ET − P_r, cm/month
Jan.	1.2	14.6	−13.4
Feb.	1.4	14.1	−12.7
Mar.	3.0	13.4	−10.4
Apr.	5.2	11.0	−5.8
May	9.8	9.6	−0.2
June	15.0	11.7	3.3
July	16.5	12.0	4.5
Aug.	16.0	6.1	9.9
Sept.	14.5	5.0	9.5
Oct.	7.2	4.5	2.7
Nov.	3.0	8.6	−5.6
Dec.	1.3	12.0	−10.7
Annual	94.1 cm	122.6 cm	−28.5 cm

The minus signs in the righthand column indicate that precipitation exceeds ET in those months and on an annual basis.

3. Determine the monthly P_w values by combining the daily P_w values and the operating days specified for each month, tabulate the results with the net ET from the table above, and determine the monthly hydraulic loadings L_w.

Month	Operating days	P_w, cm/month	Net ET, cm/month	L_w, cm/month
Jan.	10	8.4	−13.4	8.4
Feb.	12	10.1	−12.7	10.1
Mar.	15	12.6	−10.4	12.6
Apr.	30	25.2	−5.8	25.2
May	31	26.0	−0.2	26.0
June	30	25.2	3.3	28.5
July	26	21.8	4.5	26.3
Aug.	31	26.0	9.9	35.9
Sept.	25	21.0	9.5	30.5
Oct.	31	26.0	2.7	28.7
Nov.	15	12.6	−5.6	12.6
Dec.	10	8.4	−10.7	8.4
Annual	266	223.3 cm	−28.5 cm	253.2 cm

The annual hydraulic loading for wastewater on this project could be 2.5 m/year [8.20 ft/year or 61.4 gal/(ft^2 · year)]. The total liquid percolate at the site, including rainfall, would be 2.82 m (9.25 ft).

Hydraulic loading based on nitrogen limits. In many SR systems nitrogen is the LDF when protection of potable groundwater is a concern. Limitations on total nitrogen applied are based on a limiting nitrate nitrogen concentration of 10 mg/L in the receiving groundwater at the project boundary. To ensure a conservative approach the design assumption is made that the wastewater percolate will contain 10 mg/L before it mingles with the in situ groundwater. The nitrogen balance for this case is given in Eq. 7.3.

$$L_n = U + fL_n + AC_pP_w \tag{7.3}$$

where L_n = mass loading of nitrogen, kg/(ha · year) [lb/(acre · year)]
U = crop uptake, kg/(ha · year) [lb/(acre · year)]
f = fraction of applied nitrogen lost to denitrification, volatilization, and soil storage
A = conversion factor, 0.1 in SI units (2.7 USCS units)
C_p = percolate nitrogen concentration, mg/L, usually set at 10 mg/L
P_w = percolate flow, cm/year (ft/year)

Crop uptake rates can be estimated from Tables 7.5 and 7.6. The fraction lost to denitrification, volatilization, and soil storage depends primarily on the wastewater characteristics and climate. For high-strength wastewaters with BOD/nitrogen ratios of 5 or more, the f value can range from 0.5 to 0.8. Lower values apply to cold climates and higher values apply to warm climates. Recommended f values are 0.25 to 0.5 for primary municipal effluent, 0.15 to 0.25 for secondary municipal effluent, and 0.10 for advanced wastewater treatment effluents.

Equation 7.3 can be transformed and solved for P_w as shown in Eq. 7.4.

$$P_w = \frac{(1 - f)L_n - U}{0.1C_p} \tag{7.4}$$

The applied nitrogen L_n is also related to the nitrogen-limited hydraulic loading rate $L_{w,n}$, as shown in Eq. 7.5.

$$L_n = 0.1\, C_n L_{w,n} \tag{7.5}$$

where 0.1 = conversion factor (2.7 in USCS units)
C_n = nitrogen concentration in applied wastewater, mg/L
$L_{w,n}$ = hydraulic loading rate controlled by nitrogen as the LDF, cm/year (ft/year with 2.7 as conversion factor)

By combining the water balance equation, Eq. 7.1, with Eqs. 7.4 and 7.5, the value of $L_{w,n}$ can be determined:

$$L_{w,n} = \frac{C_p(P_r - \mathrm{ET}) + 10U}{(1 - f)C_n - C_p} \tag{7.6}$$

The conversion coefficient 10 is based on centimeters per year for P_r, ET, and $L_{w,n}$; the coefficient in USCS units is 0.37 when P_r, ET, and $L_{w,n}$ are expressed in feet per year.

Equation 7.6 can be used to generate monthly or annual nitrogen balances and to solve for the resulting hydraulic loading rate. The equation is conservative for design since, as stated previously, the concentration C_p in the percolate is used instead of the actual groundwater concentration at the project boundary, which may reflect further mixing and dispersion (see Chap. 3 for procedures). Example 7.2 below illustrates the use of this procedure and the method for determining the LDF for a particular project.

Example 7.2 Calculate the estimated annual nitrogen-limited hydraulic loading rate for the system described in Example 7.1. Compare this rate with the hydraulic rate as limited by soil permeability and determine the LDF for this project. The nitrogen concentration in the municipal primary effluent to be applied is 30 mg/L. Assume that orchard grass will eventually dominate the fields. From Table 7.5 assume an annual uptake U of 250 kg/(ha · year).

solution

1. Use Eq. 7.6 to determine the nitrogen-limited hydraulic loading rate. Assume C_p = 10 mg/L and since it is primary effluent, f = 0.25.

$$L_{w,n} = \frac{C_p(P_r - \text{ET}) + 10U}{(1 - f)C_n - C_p}$$

From Example 7.1, annual P_r = 122.6 cm/year, ET = 94.1 cm/year.

$$L_{w,n} = \frac{(10 \text{ mg/L})(122.6 \text{ cm/year} - 94.1 \text{ cm/year}) + 10\,[250 \text{ kg/(ha} \cdot \text{year)}]}{(1 - 0.25)(25 \text{ mg/L}) - 10 \text{ mg/L}}$$

$$= \frac{2785}{8.75}$$

$$= 318 \text{ cm/year}$$

2. Determine the LDF for this project. The maximum hydraulic loading based on soil permeability limitations is 2.5 m/year, from Example 7.1. The nitrogen-limited hydraulic loading as calculated above is 3.2 m/year. The smaller of the two is the LDF, so in this case the soil permeability controls, and the system design should be based on an annual hydraulic loading of 2.5 m/year. If this were an industrial system, it might be necessary to check other wastewater constituents as the potential LDF.

Hydraulic loading for type 2 SR systems For SR systems located in arid regions the hydraulic loading is often based on crop irrigation requirements rather than on the soil permeability since there is an economic incentive to conserve water and maximize its beneficial use. The hydraulic loading rate then depends on the irrigation requirement and any precipitation according to Eq. 7.7.

$$L_w = \text{IR} - P_r \tag{7.7}$$

where L_w = hydraulic loading rate
IR = crop irrigation requirement
P_r = precipitation

(Units must be consistent, e.g., cm/year, m/year.)

The irrigation requirement depends on the crop ET, the irrigation efficiency, and the leaching requirement. A more general form of Eq. 7.7 is given in Eq. 7.8, incorporating the leaching factor and irrigation efficiency.

$$L_w = (\mathrm{ET} - P_r)(1 + \mathrm{LR})\left(\frac{100}{E}\right) \tag{7.8}$$

where ET = crop evapotranspiration
P_r = precipitation
LR = leaching requirement
E = efficiency of the irrigation system

The leaching requirement may range from 0.05 to 0.30, depending on the crop, the amount of precipitation, and the total dissolved solids (TDS) in the wastewater. The relationship between wastewater TDS, crops, and the leaching requirement fraction is shown in Fig. 7.3. Because most wastewaters have a TDS of 400 mg/L or more, the leaching requirement is usually in the range 0.1 to 0.2.

The irrigation efficiency represents the fraction of the applied water that is accounted for in crop consumptive use or ET. The lower the efficiency, the higher the fraction of applied water that passes the root zone and percolates to deeper soil zones. For surface irrigation systems the efficiency ranges from 0.65 to 0.75; sprinkler systems usually have efficiencies of 0.7 to 0.8, and drip irrigation can achieve efficiencies of 0.9 to 0.95.

The monthly water balance is used to determine the annual hydraulic loading rate and the amount of off-season wastewater storage.

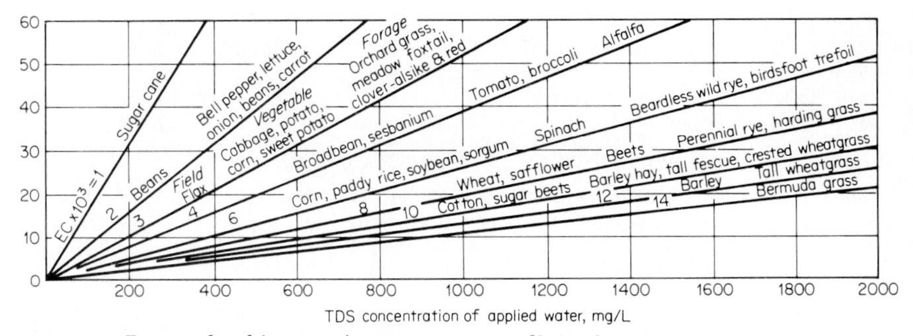

Figure 7.3 Percent leaching requirements versus salinity for various crops.

The annual hydraulic loading rate based on irrigation requirements should be checked against the nitrogen limits to determine the LDF.

Organic loading rates. Organic loading rates are not limiting for municipal SR systems and are not usually limiting for industrial SR systems. For food processing and other high-strength wastewaters, BOD loading rates often exceed 110 kg/(ha · day) [100 lb/(acre · day)] and occasionally exceed 330 kg/(ha · day) [300 lb/(acre · day)]. A list of existing systems with loadings in this range is presented in Table 7.7. These systems have successfully avoided odor problems by using adequate drying times between applications and by other management practices. Organic loading rates beyond 500 kg/(ha · day) [445 lb/(acre · day)] should generally be avoided for SR systems.

Land requirements

The land area requirements for SR systems can be significant and include the field application area plus area for roads, buffer zones, storage ponds, and preapplication treatment. The field area can be calculated by Eq. 7.9.

$$A = \frac{Q + V_s}{CL_w} \tag{7.9}$$

where A = field area, ha (acres)
Q = annual flow, m^3/year (million gal/year)
V_s = net loss or gain in stored wastewater volume due to precipitation on and evaporation and seepage from the storage pond, m^3/year (million gal/year)
C = constant = 100 (0.027 in USCS units)
L_w = design hydraulic loading rate based on the LDF, cm/year (in/year)

To determine the field area for systems with open pond storage, it is necessary to use an iterative approach because of the relationship

TABLE 7.7 BOD Loading Rates at Existing Industrial Land Application Systems[39]

Location	Wastewater type	BOD loading rate kg/(ha · day)
Almaden, McFarland, Calif.	Winery stillage	473
Bisceglia Brothers, Madera, Calif.	Winery stillage	314
Tri Valley Growers, Thornton, Calif.	Tomato	100–120
Anheuser-Busch, Houston, Tex.	Brewery	403
Anheuser-Busch, Williamsburg, Va.	Brewery	291
Ore-Ida Foods, Plover, Wisc.	Potato	215
Contadina Foods, Hanford, Calif.	Tomato	94–103

between the storage pond area and the gain or loss in volume. The procedure is described below:

1. Assume no net gain or loss in storage volume and calculate the field area.
2. Using the monthly water balance and an assumed initial depth of the storage pond, determine the net precipitation or net evaporation and/or seepage for the assumed pond. Then include this value of V_s in Eq. 7.9.
3. Solve Eq. 7.9 for the revised field area.
4. Repeat the monthly water balance using the revised field area. Adjust the surface area of the depth of the storage pond as necessary.

Storage requirements

Most SR systems require some storage for periods when cold/wet weather or crop planting/harvesting stop or reduce wastewater applications. Storage needs from cold or wet weather can be estimated from climatic data using Environmental Protection Agency (EPA) computer programs.[41] (See Fig. 2.3 for some general guidance on storage needs.) The water balance can be expanded to include columns for incremental changes in storage and cumulative storage, as shown in Example 7.3.

Application scheduling

For type 1 SR systems the applications are usually scheduled once a week for sprinkler systems and once every 2 weeks for surface irrigation. The application area is divided into subsections or sets, which are irrigated sequentially over the 1- or 2-week application cycle.

For type 2 SR systems or for nitrogen-limited SR systems with monthly nitrate limitations, the application schedule depends on the crop and the climate. The purpose is to maintain the crop in optimum growth condition by scheduling applications so that the soil moisture is depleted by no more than 30 to 50 percent.

The irrigation water requirements and leaching fraction can sometimes be achieved by a heavy preplanting application (for annual crops) or spring application (for perennial crops). The advantages of this practice are reduction in soil salinity, especially in the upper root zone, filling of the soil reservoir with water of low salinity, and reduction in the amount of leaching required during the growing season. The major disadvantage is the larger storage requirement.

Example 7.3 Determine the cumulative storage requirements for the SR system described in Example 7.1. Assume the average wastewater flow rate is 10,000 m^3/day and the design field area is 145 ha.

solution

1. Determine the wastewater volume available each month.

$$W = \frac{(10{,}000 \text{ m}^3/\text{day})(365 \text{ days/year})(100 \text{ cm/m})}{(10{,}000 \text{ m}^2/\text{ha})(145 \text{ ha})(12 \text{ months/year})} = 21.0 \text{ cm/months}$$

2. Expand the water balance table to include storage factors:

Month	L_w, cm/month	W, cm/month	Change in storage, cm/month	Cumulative storage, cm/month
Nov.	12.6	21.0	8.4	8.4
Dec.	8.4	21.0	12.6	21.0
Jan.	8.4	21.0	12.6	33.6
Feb.	10.1	21.0	10.9	44.5
Mar.	12.6	21.0	8.4	52.9
Apr.	25.2	21.0	−4.2	48.7
May	26.0	21.0	−5.0	43.7
June	28.5	21.0	−7.5	36.2
July	26.3	21.0	−5.3	30.9
Aug.	35.9	21.0	−14.9	16.0
Sept.	30.5	21.0	−9.5	6.5
Oct.	28.7	21.0	−7.7	0.0
Total	253.2 cm	252.0 cm		

3. Determine the maximum required storage volume: The peak equivalent storage from the table above is 52.9 cm/month in the month of March. The actual storage volume required is this value applied over the entire 145-ha treatment area:

$$\frac{(52.9 \text{ cm/month})(145 \text{ ha})(10{,}000 \text{ m}^2/\text{ha})}{(100 \text{ cm/m})(10{,}000 \text{ m}^3/\text{day})} = 77 \text{ days maximum storage}$$

Distribution techniques

There are three general distribution techniques—sprinkler, surface, and drip application. Suitability factors and conditions of use for these systems are presented in Table 7.8.

Sprinkler application is common to many of the more recent SR systems and all the forested systems.[10] Surface application, by either ridge-and-furrow or graded borders, is common to many of the older SR systems, especially those in the west and southwest. Drip application with wastewater effluent can lead to clogging of emitters unless proper sizing is done[22] and the effluent is screened to remove solids.[5] In addition to SS removal, the effluent should be low in iron, hydrogen sulfide, and total bacteria.[44] Drip application is an evolving technology, which will become more important for type 2 SR systems in the future.

TABLE 7.8 Suitability Factors for Wastewater Distribution Systems

Distribution technique	Suitable crops	Maximum grade, %	Minimum infiltration rate, cm/h
	Sprinkler Systems		
Solid set	No restrictions	No restrictions	0.12
Portable hand move	Orchards, pasture, grain, alfalfa	20	0.25
Side wheel roll	All crops <1 m high	10–15	0.25
Center pivot	All crops but tall trees	15	0.50
Traveling gun	Pasture, grain, field crops	15	0.75
	Surface Systems		
Graded borders (narrow, 5 m wide)	Pasture, grain, alfalfa, vineyards	7	0.75
Graded borders (wide, up to 30 m)	Pasture, grain, alfalfa, orchards	0.5–1	0.75
Straight furrows	Vegetables, row crops, orchards, vineyards	0.25	0.25
Graded contour furrows	Vegetables, row crops, orchards, vineyards	8	0.25
	Drip Systems		
	Orchards, landscape, vineyards, vegetables, nursery plants	No restrictions	0.05

Adapted from Ref. 36.

Surface runoff control

Surface runoff of applied wastewater must be controlled for most SR systems. In addition, control of stormwater runoff is usually recommended to avoid erosion problems.

For surface application systems the surface runoff of applied wastewater is known as *tailwater*. Collection of tailwater and its return to either the storage pond or the distribution system are an integral part of the design. In addition, sprinkler application systems may employ tailwater runoff control to avoid off-site discharge of applied wastewater. A usual tailwater return system consists of a collection channel or a collection sump, a pump, and a return forcemain to storage or the distribution system. Guidelines for estimating the duration of tailwater flow, the runoff volume, and the suggested maximum design volume are presented in Table 7.9.

TABLE 7.9 Recommendations for Tailwater System Design[17]

Permeability class for the soil	Permeability, cm/h	Texture range	Maximum duration of tailwater flow, %*	Estimated tailwater volume, %†	Maximum design volume, %†
Very slow to slow	0.15–0.5	Clay to clay loam	33	15	30
Slow to moderate	0.5–1.5	Clay loam to silt loam	33	25	50
Moderate to moderately rapid	1.5–15.0	Silt loams to sandy loams	75	35	70

* Maximum duration of tailwater flow as a percent of application time.
† Tailwater volumes as a percent of the application volume.

Where stormwater runoff can be significant, measures to prevent excessive erosion should be employed. Terracing of steep slopes is an accepted agricultural practice to minimize erosion. Other practices include sediment control basins, contour plowing, no-till farming, grass border strips, and stream buffer zones. Provided that wastewater application is stopped prior to a storm, the storm-induced runoff need not be collected or retained on-site.

Underdrainage

Underdrains are used in some wastewater SR systems where subsurface drainage is impeded by shallow groundwater or by a relatively impermeable layer in the soil profile. The underdrains serve to remove the water from the subsurface and thus allow wastewater application rates to continue without saturating the root zone or otherwise affecting performance.

The major consideration in adding underdrains is to maintain an unsaturated aerobic zone in the upper soil profile. An unsaturated thickness of 0.2 to 1 m (2 to 3 ft) is considered to be a minimum for effective aerobic treatment. To maintain this unsaturated thickness during operation may require either addition of underdrains or reduction in the hydraulic loading rate for a site with a subsurface restriction. The cost of installing underdrains must be compared with the cost of developing a larger land area under a lower hydraulic loading so that underdrains would not be required.

Buried plastic pipes with perforations are typically used for underdrains. Open trenches or ditches can also be used; however, if the spacing is as close as 15 m (50 ft), the ditches will consume too much land and interfere with farming operations.

Buried drains are typically about 1.8 to 2.4 m (6 to 8 ft) deep and about 10 to 15 cm (4 to 6 in) in diameter. In sandy soils typical spacings are 100 to 120 m (300 to 400 ft), with a range of 60 to 300 m (200 to

1000 ft) in practice. In clayey soils the spacings are often closer, a typical range being 15 to 30 m (50 to 100 ft). Procedures for determining drain spacing are described in Sec. 3.1.

System management

For SR systems to operate properly, the soil and crop must be well managed. The soil conditions to be managed are the infiltration rate, compaction, nutrient status, and chemical characteristics.

Infiltration rates. Soil infiltration rates can be reduced in time as a result of compaction or surface sealing. The causes include[7]:

- Compaction of the surface soils by machinery, including harvesting and cultivating equipment
- Compaction from grazing animals when the soil is wet
- A clay crust developed by water droplets or water flowing over the surface
- Clogging resulting from buildup of suspended particles, organic material, or trapped gases

The compaction or crusting can be broken up by cultivating, plowing, or other tilling operations. Minimum tillage and no-till methods minimize the compaction of soils by heavy equipment. Actively growing vegetation and decomposing residual plant material can help maintain infiltration rates at their maximum for the existing soil texture and structure. An illustration of the effect of vegetation on infiltration rates is presented in Fig. 7.4.

At sites where clay pans (hard, nearly impermeable layers) have formed and reduced the soil permeability, it may be necessary to plow to a depth of 0.6 m to 1.8 m (2 to 6 ft) or more to mix the impermeable soil layers with more permeable surface soils. Low-permeability layers in the soil profile can be modified by deep plowing.[14] This method should be considered prior to system startup to improve the initial soil permeability. Periodic deep plowing may be required if annual crops are grown, whereas less frequent plowing (at 5-year or longer intervals) may be required if perennial crops are grown.

Nutrient status. During SR system design the nutrient status of the soil should be evaluated. Sufficient nitrogen, phosphorus, and potassium are generally supplied in most municipal wastewaters. Potassium is the nutrient most likely to be deficient because it is usually present in concentrations of 10 to 15 mg/L in such waters. Some soils, particularly in the eastern United States, may be deficient in potassium. If this occurs, the vegetation cannot then function at an optimum level for nitrogen removal owing to the potassium imbalance. The need for

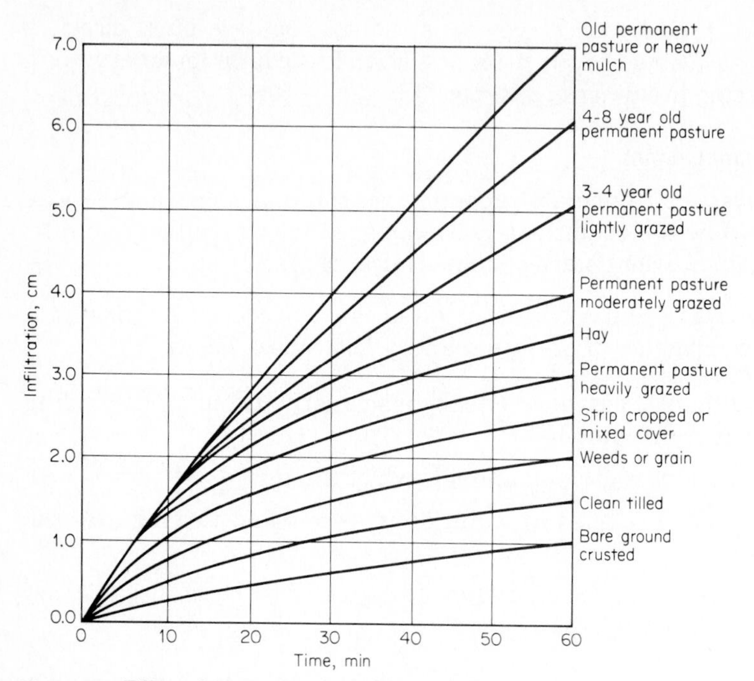

Figure 7.4 Effect of vegetation on infiltration rate.

supplemental potassium fertilizer can be estimated for this situation with Eq. 3.29.

Soil chemical characteristics. For SR systems the soil chemical characteristics of importance, in addition to soil nutrients, are pH, exchangeable sodium percentage, and salinity or electrical conductivity. The ranges of acceptable values of these parameters are presented in Table 2.15.

Soil pH can be adjusted by adding lime or gypsum (acidulating material). Exchangeable sodium can be reduced by addition of sulfur materials or calcium materials (such as gypsum), followed by leaching to remove the displaced sodium. Salinity control may require added leaching (increasing percolate flow) for type 2 SR systems.

Agricultural crop management. Annual field crops require field preparation, planting, cultivation, and harvesting. Perennial forage crops require less management, with the grass periodically harvested by cutting or grazing. Soil moisture at harvest time should be low enough so that compaction from harvesting equipment or animal hooves is minimized. The time required between the last wastewater application and harvesting depends on soil texture and drainage and the weather.

On coarse-textured soils the drying time can be as little as 3 to 4 days.[15] On fine-textured soils or where drainage is poor a drying time of 1 to 2 weeks is usually sufficient if there is no significant precipitation.

Forest crop management. Most forested SR systems are designed for existing forests,[10] and tree harvesting is not practiced for most of these systems.

If harvesting of existing forests is desired, selective harvesting and thinning are recommended. Excessive thinning can promote growth of understory vegetation and can make trees susceptible to wind throw. Thinning to develop the proper forest composition and vigor should be done prior to construction and about once every 10 years to minimize site damage and soil erosion. Wastewater applications to harvested areas should be temporarily reduced to allow the forest ecosystem to restore its treatment capacity.

System monitoring

Slow rate systems should be monitored: (1) to ensure that the desired treatment performance is being achieved; (2) to determine if any cor-

TABLE 7.10 Guidelines for Interpretation of Water Quality for Irrigation[44]

Potential problem	Units	Degree of restriction on use: None	Slight to moderate	Severe
Salinity				
EC*	dS/m, mmhos/cm	0.7	0.7–3.0	> 3.0
TDS*	mg/L	< 450	450–2000	> 2000
Permeability (based on SAR and EC)†	SAR is unitless			
SAR = 0–3	EC in dS/m	EC > 0.7	0.7–0.2	< 0.2
SAR = 3–6		EC > 1.2	1.2–0.3	< 0.3
SAR = 6–12		EC > 1.9	1.9–0.5	< 0.5
SAR = 12–20		EC > 2.9	2.9–1.3	< 1.3
SAR = 20–40		EC > 5.0	5.0–2.9	< 2.9
Specific ion toxicity				
Sodium				
Surface-applied	SAR	< 3	3–9	> 9
Sprinklers	mg/L	< 70	> 70	> 9
Chloride				
Surface-applied	mg/L	< 140	140–350	> 350
Sprinklers	mg/L	< 100	> 100	
Boron	mg/L	< 0.7	0.7–3.0	> 3.0

* EC is electrical conductivity in decisiemens per meter and TDS is total dissolved solids.
† Use SAR (sodium adsorption ratio) together with EC to evaluate potential effects on soil permeability.

rective measures are needed to protect the environment or to maintain the treatment capability; and (3) to aid in system operation. Monitoring should normally include the wastewater quality and in many cases the groundwater quality. In certain cases monitoring of the soil or vegetation may also be advisable. The values for soil chemical properties in Table 7.10 can be used in soil monitoring programs.

For type 2 SR systems the chemical properties of the wastewater to be applied should be compared with the values in Table 7.10 to determine the potential effects on crops and soils from specific constituents. For wastewaters with less than 0.7 dS/m (decisiemens per meter, or mmhos per centimeter) a leaching fraction of 15 percent would be acceptable and no other management practices would be required.[44] For clay soils it is important to consider the sodium adsorption ratio in order to avoid soil permeability problems.

7.3 Overland Flow Systems

Design objectives

Overland flow systems can be designed to achieve secondary treatment, advanced secondary treatment, or nutrient removal, depending on treatment requirements. To achieve secondary treatment the preapplication operation generally consists of fine screening, primary treatment, or equivalent treatment.

Advanced secondary treatment (15 mg/L of BOD and SS) can typically be achieved with additional preapplication treatment. Removal of nitrogen requires somewhat lower application rates than are used for BOD removal.[20] Phosphorus removal requires either pre- or postapplication treatment.

Site selection

Procedures for OF site selection are presented in Chap. 2. Overland flow is generally best suited to sites with surface soils that have permeabilities of 0.5 cm/h (0.2 in/h) or less. This low permeability is already present when the site contains fine-textured clay or clay loam soils or it can be developed by compacting a somewhat more permeable soil. Acceptable sites can also have a restrictive layer, such as hardpan or claypan at depths of 0.3 to 0.6 m (1 to 2 ft), which will result in limited deep percolation.

Preapplication treatment

Experience has shown that minimum levels of preapplication treatments can be successful when the treatment includes fine screening of municipal wastewater.[1,34,35,39] Fine screening, primary sedimentation, or a 1-day detention-time aerated pond should be considered for preapplication treatment, depending on wastewater characteristics, sludge handling concerns, and the remoteness of the site. The EPA recommends screening plus aeration (not complete-mix activated sludge) for urban locations. Removal of algal solids is difficult for OF systems.[27,46] Preapplication treatment processes such as nonaerated ponds with long detention times are not recommended for OF systems.

Climate and storage

The OF process is affected by both cold weather and rainfall. Cold weather reduces the treatment performance and usually requires storage instead of continued application, as indicated in Fig. 2.2. Rainfall also affects the performance of OF systems in terms of BOD and SS concentrations and mass discharges.[12] The effects on BOD are minimal, however, and storage is not necessary during normal rainfall events. If mass loadings of SS or strict SS concentrations must be maintained, applications may need to be curtailed during some rainfall events.

Design procedure

The empirical approach to OF design has been to select a hydraulic loading rate based on successful practice at other locations. The hydraulic loading rates have generally ranged from 2 to 10 cm/day (0.8 to 4 in/day). Recent research however, has shown that process performance is related more closely to application rate than to hydraulic loading rate.[33,34] The relationship between hydraulic loading rate and application rate is shown in Eq. 7.10.

$$L_w = \frac{qP(100 \text{ cm/m})}{Z} \tag{7.10}$$

where L_w = hydraulic loading rate, cm/day (in/day)
q = application rate per unit width of the slope, $m^3/(h \cdot m)$ [gal/(min · ft)]
P = application period, h/day
Z = slope length, m (ft)

Application rate. The relationship between application rate, slope length, and BOD removal for municipal wastewater is shown in Eq. 7.11:

$$\frac{C_z - c}{C_0} = A \exp(-KZ/q^n) \tag{7.11}$$

where C_z = effluent BOD concentration at point Z, mg/L
c = residual BOD at end of slope
= 5 mg/L
C_0 = BOD of applied wastewater, mg/L
Z = slope length, m
q = application rate, $m^3/(h \cdot m)$
K, n = empirical constants

The equation is presented graphically in Fig. 7.5 for primary effluent. It has been validated for screened raw wastewater and primary effluent, as shown in Table 7.11, but not for industrial wastewater with BOD values of 400 mg/L or more.

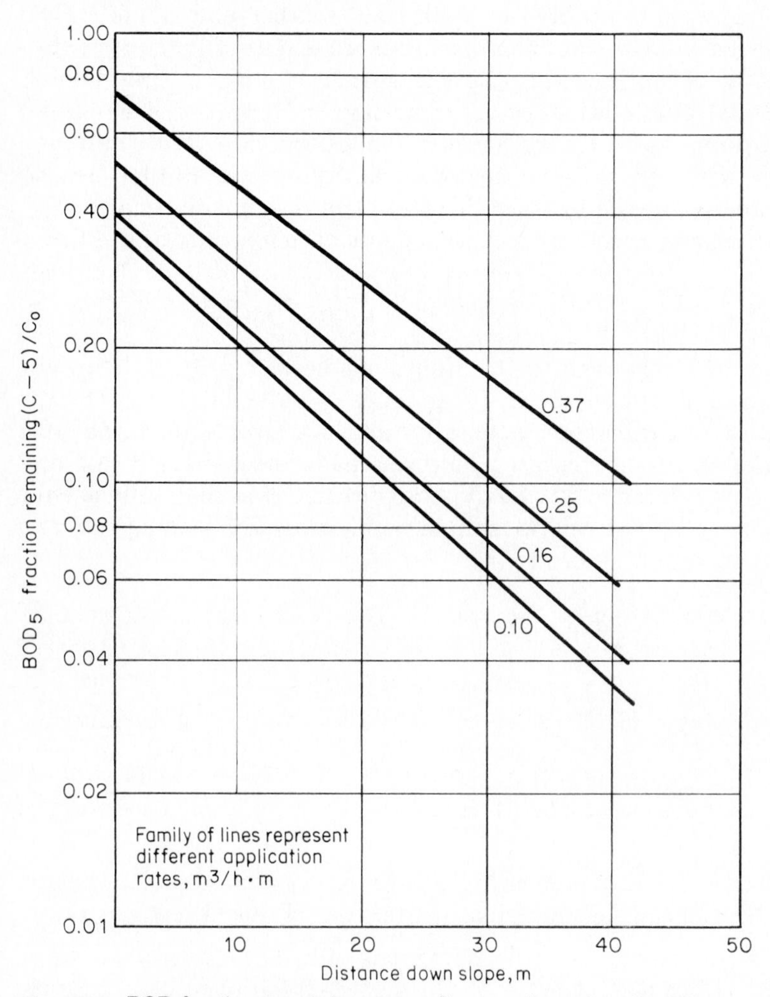

Figure 7.5 BOD fraction remaining versus distance down slope for different application rates with primary effluent.

TABLE 7.11 Comparison of Actual and Predicted OF Effluent BOD Concentrations Using Primary and Raw Wastewater[34]

Location	Applied wastewater	Application rate, $m^3/(h \cdot m)$	Slope length, m	BOD concentration, mg/L	
				Actual	Predicted
Hanover, N.H.	Primary	0.25	30.5	17	16.3
	Primary	0.37	30.5	19	17.5
	Primary	0.12	30.5	8.5	9.7
Ada, Okla.	Primary	0.10	36	8	8.2
	Raw	0.13	36	10	9.9
Easley, S.C.	Raw	0.21	53.4	23	9.6

Slope length. Treatment performance has been shown to be a function of slope length for BOD, SS, and nitrogen.[47] The higher the degree of treatment required, the longer the slope length must be. Typically, the slope length will range from 30 to 60 m (100 to 200 ft).

For surface application (gated pipe, etc.) the slope length should be 30 to 45 m (100 to 150 ft). For higher-pressure sprinkler application, which is typically used with industrial wastewaters of higher SS content, the slope length is usually 45 to 60 m (150 to 200 ft); however, the minimum length should usually be 20 m (66 ft) greater than the wetted diameter of the sprinkler application area.

Slope grade. Grades from 1 to 12 percent have been used for OF systems. At Paris, Texas the optimum range was found to be 2 to 6 percent.[28] Grades beyond 8 percent increase the risk of erosion while grades less than 1 percent increase the risk of ponding in low spots.

Application period. Application periods usually range from 6 to 12 hours each day for 5 to 7 days a week. Typically an 8-h/day period is selected in order to be compatible with normal work schedules. Occasionally OF systems can operate 24 h/day for relatively short periods.[42] The ability to oxidize ammonia is impaired with an application schedule beyond 12 h on and 12 h off.[20] The 8-h on and 16-h off schedule allows the total area to be divided into three subareas and for the system to operate 24 h/day when required.

Organic loading rates. Organic loading rates for OF are limited to the oxygen transfer efficiency into and through the thin film of wastewater (usually 0.5 cm deep or less) on the slope. The limiting rate appears to be about 100 kg/(ha · day) [89 lb/(acre · day)] to avoid excessive anaerobic conditions on the slope. The organic loading rate can be calculated from Eq. 7.12.

$$L_{BOD} = 0.1 L_w C_0 \quad (7.12)$$

where L_{BOD} = BOD loading rate, kg/(ha · day) [lb/(acre · day)]
0.1 = conversion factor (0.225 in USCS units)
L_w = hydraulic loading, cm/day (in/day)
= $qPWm/Z$
q = application rate, $m^3/(h \cdot m)$ [gal/(min · ft)]
P = application period, h
W = width of application slope, m (ft)
Z = length of application slope, m (ft)
m = conversion factor
= 100 cm/m (96.3 for USCS units)
C_0 = BOD of applied wastewater, mg/L

When the BOD of the applied wastewater exceeds about 800 mg/L, the oxygen transfer capacity of OF systems becomes limiting. To overcome this constraint one industrial wastewater system used an effluent recycle system. The raw wastewater with a BOD of 1700 to 1800 mg/L, was diluted 1:1 and 3:1 with system runoff in a pilot study.[26] The resultant BOD removal was 97 percent at a BOD loading rate of 56 kg/(ha · day) [50 lb/(acre · day)]. The concept has been proved on a full-scale basis at Rosenberg, Texas.

Suspended solids loadings. With the exception of algae, wastewater solids will generally not be limiting in OF system designs. Suspended solids are effectively removed on the slopes by sedimentation and filtration because of the low velocity and the shallow depth of flow. When surface application methods are used, most of the SS will be removed within the first few meters of the application point. This can create sludge problems with (industrial) wastewaters of high SS content, and the use of sprinklers for more uniform distribution is recommended.

Nitrogen removal. Nitrogen removal is dependent on adequate BOD/nitrogen ratios, adequate detention time (low application rates and long slopes), and temperature. Nitrification and denitrification account for most of the nitrogen removal.[29] Soil temperatures below 4°C will limit the nitrification reaction. Denitrification appears to be most effective when screened raw or primary effluent is applied because of the high BOD/nitrogen ratio. The application of municipal effluents with nitrate instead of ammonia to OF slopes results in very little nitrogen removal.[24] Apparently, the initial retention of ammonia in the soil by adsorption is an important first step prior to nitrification and denitrification.

Up to 90 percent removal of ammonia was reported at 0.10 $m^3/(h \cdot m)$ [0.13 gal/(min · ft)] at the OF system in Davis, California.[20] Slope lengths of 45 to 60 m (150 to 200 ft) may be required to achieve this level of ammonia removal.[34]

Land requirements

The field area required for OF depends on the flow, the application rate, the slope length, and the period of application, as shown in Eq. 7.13, which assumes no seasonal wastewater storage.

$$A_s = \frac{QZ}{qPC} \tag{7.13}$$

where A_s = field (surface) area required, ha (acres)
Q = wastewater flow rate, m^3/day (gal/min)
Z = slope length, m (ft)
q = application rate, $m^3/(h \cdot m)$ [gal/(min · ft)]
P = period of application, h
C = conversion factor
= 10,000 m^2/ha (726 USCS units)

If wastewater storage is a project requirement, the field area is determined with Eq. 7.14.

$$A_s = \frac{365Q + V_s}{DL_wC'} \tag{7.14}$$

where V_s = net loss or gain in storage volume due to precipitation, evaporation, and seepage, m^3/year (ft^3/year)
D = number of operating days per year
L_w = design hydraulic loading (see Eq. 7.12 for definition), cm/day (in/day)
C' = conversion factor
= 100 (metric), 3630 (USCS)

If the organic loading rate is limiting, the field area can be calculated by Eq. 7.15.

$$A_s = \frac{C_0 C'' Q_a}{L_{\mathrm{LBOD}}} \tag{7.15}$$

where A_s = field area, ha (acres)
C_0 = BOD of applied wastewater, mg/L
C'' = conversion factor
= 0.1 (metric units) [6.24×10^{-5} (USCS units)]
Q_a = design flow rate to the OF site, m^3/day (ft^3/day)
L_{LBOD} = limiting BOD loading rate
= 100 kg/(ha · day) [89 lb/(acre · day)]

Example 7.4 Determine the field area requirements for a municipal OF system to treat 4000 m³/day. The primary effluent has a BOD of 150 mg/L and the effluent discharge BOD limit is 30 mg/L. Assume 20 days of storage.

solution

1. Compute the required removal ratio.

$$\frac{C_z - 5}{C_0} = \frac{30 - 5}{150} = 0.17$$

2. Using Fig. 7.5, enter the graph at a BOD remaining fraction of 0.17 and proceed to the maximum application rate, or 0.37 m³/(h · m). The resultant slope length is 30 m.
3. Select an application period of 8 h/day.
4. Using a safety factor of 1.5, compute the design application rate q.

$$q = \frac{0.37}{1.5} = 0.25\ \text{m}^3/(\text{h}\cdot\text{m})$$

5. Calculate the hydraulic loading rate.

$$L_w = \frac{qP}{Z}$$

$$= \frac{(0.25)(8)}{30} = 0.067\ \text{m/day}$$

6. Calculate the number of operating days.

365 − 20 = 345 days/year

7. Assuming that the seepage and evaporation from the storage pond offset the precipitation, calculate the field area.

$$A_s = \frac{(4000\ \text{m}^3/\text{day})(365) + 0}{(345)(0.067\ \text{m/day})(10{,}000\ \text{m}^2/\text{ha})}$$

$$= 6.3\ \text{ha}$$

The application rate used for design of municipal OF systems depends on the LDF, the climate, and to some extent, the preapplication treatment level. Suggested application rates and hydraulic loading rates are presented in Table 7.12.

TABLE 7.12 Application Rates Suggested for Overland Flow Design

Pre-application treatment	Stringent requirements and cold climates		Moderate requirements and climates		Least stringent requirements and warm climates	
	m³/(h · m)*	cm/day	m³/(h · m)	cm/day	m³/(h · m)	cm/day
Screening/ primary	0.07–0.1	2	0.16–0.25	3–5	0.25–0.37	5–7
Aerated cell (1-day detention)	0.08–0.1	2	0.16–0.33	3–6	0.33–0.4	6–8
Secondary	0.16–0.2	4	0.2–0.33	4–6	0.33–0.4	6–8

* m³/(h · m) × 1.34 = gal/(min · ft).

Winter operation. Cold weather storage requirements are not well defined for OF systems because of the limited operating experience available. Figure 2.2 presents some general guidelines based on geographical location. Operations can continue at soil temperatures near 0°C and with surface application systems beneath a snow cover. Wastewater applications should cease when an ice cover forms on the slope. Operation of sprinkler systems can be very difficult at air temperatures below freezing. In locations where night-time temperatures fall below 0°C (32°F) but daytime temperatures exceed 2°C (36°F), a day-only operation may be chosen in which all the field area is used within 10 to 12 hours.

Storage area. The winter or operating storage pond should be located off-line so that it only contains wastewater for the minimum time. The pond should be drained as soon as possible when application permits so that algae growth is minimized. As indicated in Chap. 4, the storage ponds are generally more than 3 m (10 ft) deep to minimize the land area required.

Vegetation selection

The grass used on the OF slopes is important for ability to provide a support medium for microorganisms, to minimize erosion, and to take up nitrogen and phosphorus. The crop is not intended to be marketed unless the other important functions are fulfilled. The grass should be a perennial species, have high moisture tolerance, have a long growing season, and be suited to the local climatic conditions. A list of grasses commonly used in OF systems is presented in Table 7.13.

A mixture of grasses is recommended for most sites. The mixture should include warm season and cool season species, as well as sod formers and bunch grasses. Some grasses such as reed canary grass are slow to become established and require a nurse crop such as annual rye grass for ground cover during the early part of the first operational season. Rye grass will not usually last more than a few years on the slope.

Local agricultural advisers should be consulted to select the grasses listed in Table 7.13. Pure stands of grasses such as Johnson grass, yellow foxtail, and most grass species that have a single seed stalk should be avoided.

Most of the grasses for OF are established by seeding, although coastal Bermuda and other improved Bermuda grasses need to be sprigged. Details on seeding and establishment of OF grasses are provided in Ref. 42. Hydroseeding may also be used if the range of the

TABLE 7.13 Perennial Grasses Suitable for OF Systems[42]

Common name	Rooting characteristics	Growing height, cm
Cool Season Grasses		
Reed canary	Sod	120–210
Tall fescue	Bunch	90–120
Redtop	Sod	60–90
Kentucky bluegrass	Sod	30–75
Orchard grass	Bunch	15–60
Warm Season Grasses		
Common Bermuda	Sod	30–45
Coastal Bermuda*	Sod	30–60
Dallis grass	Bunch	60–120
Bahia	Sod	60–120

* Includes other improved Bermuda grass varieties.

distributor is adequate to provide coverage of the slopes without vehicular travel on the seedbed.

The grasses should be watered with a portable sprinkler system as soon as seeding is complete. Short, frequent waterings with fresh water are preferred, and no runoff should be allowed. The permanent wastewater distribution system should not be used until the grass is well established to avoid erosion of the bare soil. The first grass cuttings should be allowed to remain on the slope to help build up the organic mat, provided that the clippings are relatively short [15 cm (6 in)]. An OF system may take 3 to 4 months of acclimation and start-up before full treatment capability is established.

Distribution system

Municipal wastewater can be surface-applied to OF systems; however, industrial wastewater should be sprinkler-applied. Surface application using gated pipe offers lower energy demand and avoids aerosol generation. Slide gates at 0.6-m (2-ft) spacings are recommended over screw-adjusted orifices. Pipe lengths of 100 m (300 ft) or more require in-line valves to allow adequate flow control and isolation of pipe segments for separate operation.

Sprinkler distribution is recommended for wastewater with BOD or SS levels of 300 mg/L or more. Impact sprinklers located about one-third of the way down the slope are generally used. Wind speed and direction must be considered in locating sprinklers and in determining the overlap in the spacing between sprinklers.

Slope design and construction

Even naturally occurring slopes must be regraded or reshaped to ensure a smooth surface for uniform flow of water down the slope. Design methods are detailed in Ref. 42.

Where extensive cut-and-fill operations are used, settling of the fill sections may occur after rough grading. If such precipitation is absent, it may be necessary to water the slope and correct for any settling that occurs thereafter. Existing depressions should be brought level with adjacent areas, with 15 cm (6 in) lifts compacted to the density of the adjacent undisturbed soils.

After the slopes have been formed in the rough grading operation, a heavy disk should be used to break up the large clods of soil. A land plane should then be used to smooth out the slope. Typically, a grade tolerance of 1.5 cm (0.05 ft) from the final elevations can be achieved with three passes of the land plane.

Runoff collection

Treated effluent is usually collected in open drainage channels at the toe of the slope. There may be cases in which the drainage channel is lined or converted into a pipeline. Earthen collection channels are usually vegetated with the same grass species as the slopes and are graded to prevent erosion. Side slopes on V-type channels should not exceed 4:1.

Unless upstream drainage is channeled around the OF site, the drainage channels and discharge structures should be designed for the discharge from the entire site, not just the OF slopes. Drainage channels should have adequate capacity to contain the peak rate of runoff from a 25-year, 24-h frequency storm, with 0.1 m (4 in) of freeboard as a minimum.

Recycle

For high-strength industrial wastewater, a recycle operation to blend treated runoff with influent wastewater may be practical. The collection system should include a sump from which the treated runoff can be pumped back to the distribution system. Stormwater runoff should be allowed to bypass the recycle sump and be discharged directly.

System management and monitoring

Overland flow systems require minimal management. The grass should be cut two or three times per year. The cuttings can be left on the slope unless nitrogen removal is a project expectation. If the cut grass is too tall, i.e., higher than 30 cm (12 in), it should be removed so as not to smother the new growth.

Slopes must be sufficiently dry prior to mowing that no ruts or depressions are formed. The drying time may range from a few days to 2 weeks, depending on the soil and climatic conditions.

Weeds and native grasses will often begin growing on the slopes. They are of concern only if they invade and replace the planted species, especially if the weeds or native grasses are annual species and replace the perennial grasses intended for the system. In some cases burning of the weeds or disking and reseeding of grass may be necessary.

System monitoring includes influent and runoff quality and flow, groundwater quality and levels, surface water quality, and soil and vegetation characteristics. Only two groundwater monitoring wells are usually needed unless groundwater levels are relatively high or permeable soils are used. Measurements of surface water quality above and below the final point of system discharge are usually required. Monitoring of soils and vegetation is similar to that for SR systems.

7.4 Rapid Infiltration Systems

Process design procedures and examples for RI systems are presented in this section. Details on the critically important site selection and infiltration testing are presented in Chap. 2, and groundwater management and drainage fundamentals are presented in Chap. 3.

Design objectives

The design objectives for RI systems can include treatment followed by:

- Recharge of streams by interception of groundwater
- Recovery of water by wells or underdrains, with subsequent reuse or discharge
- Groundwater recharge
- Temporary storage of renovated water in the aquifer

The typical use of RI involves the indirect recharge of adjacent surface waters, with occasional systems having underdrains.[8]

Design procedure

The process design procedure for RI is outlined in Table 7.14. The first step is to adequately characterize the site, as described in Chaps. 2 and 3. The potential infiltration rate is based on the infiltration testing

TABLE 7.14 Procedures for Rapid Infiltration Process Design

Step	Description
1	Determine the potential infiltration rate.
2	Predict the hydraulic pathway.
3	Determine treatment requirements.
4	Select preapplication treatment level.
5	Calculate the annual hydraulic loading rate.
6	Calculate field area.
7	Check for groundwater mounding.
8	Select the final hydraulic loading cycle.
9	Determine the application rate.
10	Determine the number of basins.
11	Determine the monitoring requirements.

performed at the site. Generally, the mean of a number of steady-state test results is used as the basis for this determination.

The site work should also include numerous backhoe pits and soil borings to determine the subsurface lithology. Based on the soil profile and hydrogeology of the site, the hydraulic pathway of the treated percolate can be predicted. The percolate will flow to groundwater unless a subsurface layer impedes the vertical flow.

Treatment performance

The treatment performance of RI systems generally improves as the hydraulic loading rate decreases. The improvement in BOD and SS removal is only slight for loading rates less than 100 m/year (330 ft/year). The improvement is more marked for nitrogen and phosphorus removal, and if removal of these constituents is required, the loading rate for them should be considered. In addition, organic loading rates for industrial wastewaters should be checked.

Nitrification

Ammonia removal by nitrification can be readily accomplished by RI. Although nitrification is affected by temperature, the recent experience at Boulder, Colorado has shown that nitrification can occur even at a temperature of 4°C (40°F).[29] Reducing the application rates in the cold periods will compensate for reduced rates of nitrification and will also allow more of the applied ammonia to be absorbed by the soil.

Nitrification rates as high as 67 kg/(ha · day) [60 lb/(acre · day)] have been reported.[41] Ammonia loading rates should not exceed this value if nitrification is a treatment objective. The loading cycle for nitrification should consist of 1 to 3 days of flooding followed by 5 to 10 days of drying to restore aerobic conditions in the near-surface soil profile.

Nitrogen removal

Nitrogen removal by denitrification in RI systems requires both adequate detention time and adequate organic carbon to be effective. Generally, primary effluent with a BOD/nitrogen ratio of 3:1 or higher provides adequate organic carbon for the denitrification reaction. Secondary effluent, however, contains insufficient organic carbon to achieve more than about 50 percent nitrogen removal. To improve nitrogen removal with secondary effluent it is necessary to flood the basins for as long as 7 to 9 days, followed by 12 to 15 days of drying.

Nitrogen removal is also related to the infiltration rate, as demonstrated with secondary effluent at Phoenix, Arizona.[21] As shown in Figure 7.6, nitrogen removal increased in the column studies from about 30 to 80 percent when the infiltration rate decreased from 30 to 15 cm/day (12 to 6 in/day).

To achieve 80 percent nitrogen removal in RI systems it appears that the following maximum infiltration rates should not be exceeded:

Primary effluent:	20 cm/day (8 in/day)
Secondary effluent:	15 cm/day (6 in/day)

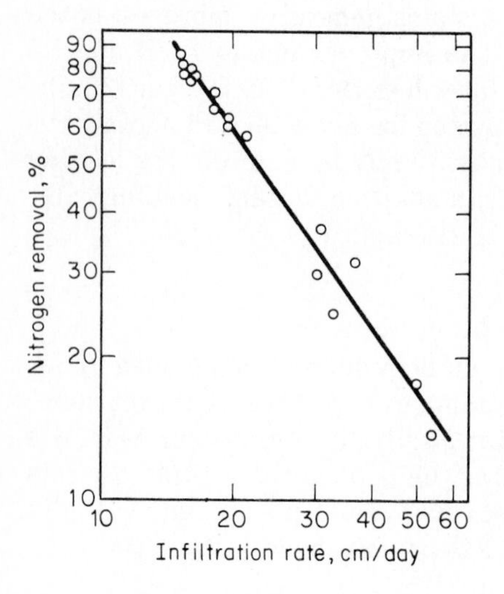

Figure 7.6 Effect of infiltration rate on nitrogen removal.

The design procedure for nitrogen removal in RI systems is as follows:

1. Calculate the mass of ammonia nitrogen that can be adsorbed as ammonium based on the cation-exchange capacity of the soil.
2. Based on the wastewater ammonia concentration and daily application rate (hydraulic loading rate in meters per cycle divided by the number of days of application per cycle), calculate the length of the loading period that can be used without exceeding the mass loading from step 1.
3. Compare the ammonia and organic nitrogen (combined) loading rate with the 67 kg/(ha · day) [60 lb/(acre · day)] criterion to ensure that nitrification can be expected.
4. Based on the ammonium adsorption capacity (step 1), establish the length of the acceptable loading period. The drying period to achieve nitrification and denitrification should be selected from Table 7.15.
5. Balance the nitrate nitrogen produced against the BOD applied to ensure an adequate BOD/nitrogen ratio.
6. Reduce the infiltration rate as necessary to produce the nitrogen removal needed. The infiltration rate can be reduced by lowering the depth of flooding, by incorporation of finer-textured soils into the soils of the basin's infiltration surface, or by compaction of the soil.

Nitrification and denitrification will be reduced in cold weather. Continued application during freezing conditions will result in accumulation of ammonium in the soil profile. In the spring, with the onset

TABLE 7.15 Suggested Loading Cycles for Rapid Infiltration[41]

Objective	Wastewater type	Season	Application period, days*	Drying period, days
Maximize infiltration rate	Primary	Summer	1–2	5–7
		Winter	1–2	7–12
	Secondary	Summer	1–3	4–5
		Winter	1–3	5–10
Maximize nitrification	Primary	Summer	1–2	5–7
		Winter	1–2	7–12
	Secondary	Summer	1–3	4–5
		Winter	1–3	5–10
Maximize nitrogen removal	Primary	Summer	1–2	10–14
		Winter	1–2	12–16
	Secondary	Summer	7–9	10–15
		Winter	9–12	12–16

* Regardless of season or objective, application periods for primary effluent should be limited to 1 to 2 days to prevent excessive soil clogging.

of warm weather, nitrification of the adsorbed ammonium will result in a high concentration of nitrate nitrogen in the percolate until the denitrifying bacteria become active and can begin to assimilate the available nitrate. For systems requiring nitrogen removal throughout the year, some winter storage may be required at cold climate locations.

Phosphorus removal

The phosphorus removal capability of an RI system can be estimated by using Eq. 3.28. The detention time is critical and is a function of the percolation rate through the soil profile and of the flow distance. If the flow distance is insufficient, reevaluation of the selected site or use of some other method for phosphorus removal may be necessary. The infiltration rate can be reduced by compacting the soil or by lowering the depth of wastewater application. These changes can influence the residence time in the near-surface soils but will have little effect on the deeper natural profile.

Phosphorus removal at existing RI systems is better than would be predicted from Eq. 3.28 since the flow is actually unsaturated in the near-surface profile, and the equation assumes "worst case" saturated flow and therefore a shorter detention time. Removals at five RI systems are presented in Table 7.16, along with the travel distance to the sampling point. As indicated in Chap. 3, phosphorus adsorption tests can be conducted with site-specific soils if phosphorus removal is critical.

Preapplication treatment

Once the treatment requirements have been established and the capacity of the RI system for treatment determined, the level of preapplication treatment can be selected. For municipal wastewater the appropriate level of preapplication treatment is typically primary sedimentation. An equivalent level of SS removal can be achieved with a

TABLE 7.16 Phosphorus Removal Data for Selected RI Systems[8]

Location	Applied concentration, mg/L	Distance to sample point, m	Percolate concentration, mg/L	Removal, %
Calumet, Mich.	3.5	1700	0.03	99
Dan Region, Israel	2.1	150	0.03	99
Ft. Devens, Mass.	9.0	45	0.10	99
Lake George, N.Y.	2.1	600	0.014	99
Phoenix, Ariz.	5.5	30	0.37	93

short-detention-time pond. The advantage of the pond is the saving in sludge handling, treatment, and disposal.

Long-detention-time ponds of the type described in Chap. 4 are generally not appropriate for use ahead of RI basins. The algae produced in oxidation ponds will significantly reduce the infiltration rates in RI systems. Biological treatment may be cost-effective prior to the RI component in urban settings.

Hydraulic loading rates

The design hydraulic loading for an RI system is usually limited by the hydraulic characteristics of the soil profile because of the intent to apply large volumes of wastewater to a relatively small surface area. For some systems, however, the LDF may be the nitrogen or BOD loading rate.

The *potential infiltration rate* is the steady-state rate at which wastewater would be expected to infiltrate into the soil for a significant period of time. To account for the cyclical nature of RI loadings, the normal variability of site conditions, and the limitations of the relatively small-scale field-test procedures, a small percentage of the potential infiltration rate is used for design purposes to calculate the annual hydraulic loading rate. If the basin infiltration test is used and if site conditions are generally uniform, the safety factor can be 7 to 15 percent. For the much smaller-scale cylinder infiltrometer tests or air entry permeameter measurements, the safety factor should only be 2 to 4 percent because of the smaller zone of influence of these test procedures. The selection of a specific safety factor depends on the number and type of field measurements and their variability and on the uniformity of soil conditions. When a large number of field tests are conducted, the results are not widely variable, and soil conditions are generally uniform over the site, the larger safety factors can be used, as shown in Table 7.17.

TABLE 7.17 Design Safety Factor for RI Hydraulic Loading Determination

	Safety factor	
Procedure	Condition A*	Condition B†
Basin flooding test	7–10	10–15
Cylinder infiltrometer, air entry permeameter and similar small scale tests	2	4

* Conservative range, use with variable field data or site conditions.

† Less conservative value, appropriate when variability of test results and site conditions are low.

Example 7.5 Determine the hydraulic loading rate and the application rate for an RI system using primary effluent and with the design intention of maximizing infiltration rates. The measured infiltration rate using the basin flooding test is 4 cm/h. Few field tests were conducted, and the results were quite variable.

solution

1. Compute the annual potential infiltration rate using the 4 cm/h steady-state field test results.

$$\frac{(4\ \text{cm/h})(24\ \text{h/day})(365\ \text{days/year})}{100\ \text{cm/m}} = 350.4\ \text{m/year}$$

This is the potential annual rate at which water could infiltrate on a continuous year-round basis if the soil profile were uniform and aerobic and clogging did not occur. Since none of these requirements are likely to prevail, it is necessary to use a safety factor from Table 7.17 for design.

2. Calculate the annual hydraulic loading. Since the test results were variable, select a conservative percentage from Table 7.17; the range 7 to 10 percent is appropriate. Use 8.5 percent as the midpoint of the range.

$$L_w = (0.085)(350.4\ \text{m/year}) = 29.8\ \text{m/year}$$

This annual hydraulic loading is the volume of water that might be applied to the site on a year-round basis if there were no downtime for seasonal restrictions, maintenance, or special operational cycles involving very short wetting periods followed by a very long drying period. For the typical case this annual hydraulic loading can be determined prior to selection of the wastewater loading cycle, and it will be compatible with all of the typical loading cycles given in Table 7.15. If year-round operation is not intended, it is necessary to proportionally reduce the annual hydraulic loading rate for the nonoperational periods.

3. Determine the application rate. Since the design intent is to maximize infiltration rates, select a loading cycle of 2 days application and 12 days drying from Table 7.15, since winter conditions will control.

$$\text{Cycles per year} = \frac{365\ \text{days/year}}{14\ \text{days/cycle}} = 26\ \text{cycles/year}$$

$$\text{Application rate} = \frac{L_w}{\text{cycles/year}} = \frac{29.8\ \text{m/year}}{26\ \text{cycles/year}}$$

$$= 1.15\ \text{m/cycle}$$

$$\text{Daily application rate} = \frac{1.15\ \text{m/cycle}}{2\ \text{day/cycle}} = 0.58\ \text{m/day}$$

This is the average daily application rate during the 2-day application period at the start of each cycle. This application rate is less than the steady-state infiltration rate (4 cm/h × 24 h/day ÷ 100 cm/m = 0.96 m/day), so all the applied water should infiltrate soon after the 2-day application period is complete, leaving the balance of the remaining 12 days for drying of the basins. Some clogging may occur after long-term operation, and eventually maintenance will be required. Assume, for example, that the infiltration rate in this case is reduced to about 25 percent of the measured steady-state rate (0.96 m/day) during the second day of wastewater application, and calculate the total time required to infiltrate the applied water.

$$t = 1 + \frac{0.575\ \text{m/day}}{(0.25)(0.96\ \text{m/day})} = 3.4\ \text{days}$$

This would still leave almost 11 days for drying, which is adequate for this type of system.

Organic loading rates

For RI systems the exact limit for organic loading has not been developed. Experience with winery wastewater has shown that BOD loading rates above 670 kg/(ha · day) [600 lb/(acre · day)] can lead to odor problems.[9] A primary effluent loaded at 8 cm/day and with a 150-mg/L BOD level would generate 120 kg/(ha · day) [107 lb/(acre · day)] of BOD loading, which should not create odor problems.

Land requirements

Land requirements include the RI basins, access roads, preapplication treatment, plus any buffer zone or area for future expansion. Land for the required bottom area in the RI basins is calculated by Eq. 7.16. Additional area must be added for dikes or berms and access ramps.

$$A = \frac{CQ(365 \text{ days/year})}{L_w} \tag{7.16}$$

where A = application area, ha (acres)
C = conversion factor
= 10^{-4} ha/m², [3.06×10^{-6} (acre · ft/gal)]
Q = average wastewater flow, m³/day (million gal/day)
L_w = annual loading rate, m/year (ft/year)

The land areas required for the nitrification and organic loading rate criteria should be checked, if applicable. If larger areas are required for these loading rates, the largest area should be used.

Number of basin sets. The number of RI basins or sets of basins depends on the hydraulic loading cycle and the topography of the site. The decision on the number of basins and the number to be flooded at one time affects the distribution system hydraulics. The minimum number of basins required for a continuous flow of wastewater is presented in Table 7.18.

Application rate. The application rate is determined by the annual loading rate and the hydraulic loading cycle. To determine the application rate, divide the annual loading rate by the number of loading cycles per year and then divide the loading per cycle by the application period, as shown in Example 7.5.

Groundwater mounding. Some temporary mounding of groundwater beneath the basin due to percolate flow is acceptable provided that it does not interfere with infiltration at the basin surface and provided

TABLE 7.18 Minimum Number of RI Basins Required for Continuous Wastewater Flow and Year-Round Application[41]

Application period, days	Drying period, days	Minimum number of basins
1	5–7	6–8
2	5–7	4–5
1	7–12	8–13
2	7–12	5–7
1	4–5	5–6
2	4–5	3–4
3	4–5	3
1	5–10	6–11
2	5–10	4–6
3	5–10	3–5
1	10–14	11–15
2	10–14	6–8
1	12–16	13–17
2	12–16	7–9
7	10–15	3–4
8	10–15	3
9	10–15	3
7	12–16	3–4
8	12–16	3
9	12–16	3

that it dissipates quickly enough to allow for aerobic restoration of the near-surface soil profile. Groundwater mounding equations and nomographs in Chap. 3 can be used to determine if the groundwater mound will interfere with RI operations. If mounding will interfere, there are several options to consider, including adjustment of the flooding and drying cycle, alternation of the operations pattern of basin usage to minimize mounding, reduction in the loading rate, and addition of underdrains.[42] Basins can be separated within the site, if practical, by placing preapplication treatment or administrative facilities between sets of basins.

Basin construction

Care should be taken in constructing RI basins so as not to compact the infiltration surfaces. The design should, if possible, avoid basin construction on backfilled materials. This is because standard construction equipment will tend to compact the fill material and may do irreparable damage to the hydraulic properties of the soil.

The permeability of soil at a given field density can vary depending on the moisture content of the soil at the time of compaction. If a soil containing significant clay content is worked and compacted on the "wet" side of the optimum moisture content, the permeability could be lower by an order of magnitude than the value that might be expected if the soil were placed with the moisture content on the "dry" side of optimum. On the basis of experience with failure of RI systems built on fill material,[30] the following recommendations should be followed if the use of backfill is a project necessity:

1. If fill is to be used, at least one basin flooding test should be conducted in a "test" fill, which should be constructed with the same equipment intended for full-scale construction. The width of the test fill area should be twice the diameter of the test basin area, and the depth of the fill should be equal to the final design fill depth or 1.5 m (5 ft), whichever is less. The design hydraulic loading rate for these fill areas should be based on the results of these tests.
2. Placement of any fill in the infiltration area should only be conducted when the soil moisture content is on the dry side of optimum.
3. Clayey sands with a clay content of 10 percent or more are unsuitable for use as fill material in RI basins.
4. The construction sequence to be followed using dry soils, is:
   a. Cut or fill to the specified elevation.
   b. Fine grade to the specified tolerance.
   c. Rip the basin bottoms to a depth of 0.6 to 1 m (2 to 3 ft) in two perpendicular directions.
   d. Disk the surface to break up consolidated material.

The dikes around the RI basins need not be very high; 1 m (3 ft) or less is usually adequate. Tall dikes increase the construction cost, increase the potential for erosion, and compound the problem of access to the basins. Erosion control of dike soils is important during construction to ensure that fine-textured materials are not washed onto the infiltration surfaces in the basins. The use of silt fences or other barriers is recommended until grass is established. A ramp into each basin for maintenance equipment is essential.

Winter operation in cold climates

Rapid infiltration systems can operate successfully on a year-round basis, as has been shown in such cold climates as those of Idaho, Montana, South Dakota, Michigan, Wisconsin, and New York State. Proper thermal protection for the pipes, valves, and pump stations is essential. Ice formation on or in the upper soil profile is the critical problem to avoid during winter operation of an RI system. Approaches that can be used successfully include:

1. Ridge and furrow surface application combined with a floating ice sheet. The ice gives thermal protection to the soil and rests on the ridge tops as the wastewater infiltrates the furrows.

2. Induction of snow drifting with snow fences in the basins followed by flooding beneath the snow layer.
3. Design of one or more basins for continuous loading during extreme conditions. These basins would be taken off-line and "rested" for an extended period during the following summer.
4. Retention of the available heat in the wastewater by using preapplication treatment processes with short detention times.

System management

It is essential that RI basins be operated on an intermittent basis. The scheduling of the application period and the drying period is an important operational task. The length of time required for each basin to dry should be recorded each cycle. An increase in the intended drying time can be an indication of the need for basin maintenance.

Periodic basin maintenance may include scarification, scraping, or disking of the basin surface. Equipment traveling across the basins should be minimized and only allowed on the basins when the soil is dry. Any thick deposits of soil fines or organic material should be removed. If grass is grown, it should be mowed or should be burned at least once per year, preferably just before winter.

System monitoring

Monitoring should be conducted to provide data for system management or adjustments and to comply with regulatory requirements; it should cover the applied wastewater, groundwater quality, and groundwater levels. Groundwater wells should be placed both up gradient and down gradient from the application area. Details of groundwater monitoring wells can be found in Ref. 32.

REFERENCES

1. Abernathy, A. R., J. Zirschky, and M. B. Borup: "Overland Flow Wastewater Treatment at Easley, S.C." *J. Water Pollution Control Fed.*, 57(4):291–299, April 1985.
2. Bishop, P. L., and H. S. Logsdon: "Rejuvenation of Failed Soil Absorption Systems," *J. Environ. Eng. Div.*, ASCE, 107(EE1):47, 1981.
3. Bouwer, H., and R. C. Rice: "Renovation of Wastewater at the 23rd Avenue Rapid Infiltration Project," *J. Water Pollution Control Fed.*, 56(1):76–83, January 1984.
4. Broadbent, F. E., and H. M. Reisenauer: "Fate of Wastewater Constituents in Soil and Groundwater: Nitrogen and Phosphorus," Pettygrove and Asano (eds.), *Irrigation with Reclaimed Municipal Wastewater—a Guidance Manual,* California State Water Resources Control Board, July 1984.
5. Cadiou, A., and L. Lesavre: "Drip Irrigation with Municipal Sewage, Clogging of the Distributors," *Proc. Third Water Reuse Symp.*, San Diego, Calif., Aug. 26–31, 1984.

6. California Fertilizer Association: *Western Fertilizer Handbook,* Interstate Printers & Publishers Inc., Danville, Ill., 7th Ed., 1985.
7. Crites, R. W.: "Site Characteristics," *Irrigation with Reclaimed Municipal Wastewater—Guidance Manual,* California State Water Resources Control Board, July 1984.
8. Crites, R. W.: "Nitrogen Removal in Rapid Infiltration Systems," *J. Environ. Eng. Div.,* ASCE, 111(6):865–873, December 1985.
9. Crites, R. W., and R. C. Fehrmann: "Land Application of Winery Stillage Wastes," *Proc. Third Ann. Madison Conf. Applied Res. Municipal Industrial Waste,* Sept. 10–12, 1980, pp. 12–21.
10. Crites, R. W., and S. C. Reed: "Technology and Costs of Wastewater Application to Forest Systems," *Proc. Forest Land Applications Symp.,* Seattle, Wash., 1986.
11. Doorenbos, J., and W. O. Pruitt: "Crop Water Requirements," *FAO Irrigation and Drainage Paper No. 24,* United Nations Food & Agr. Org., Rome, 1977.
12. Figueiredo, R. F., R. G. Smith, and E. D. Schroeder: "Rainfall and Overland Flow Performance," *J. Environ. Eng. Div.,* ASCE, 110(3):678–694, June 1984.
13. Folsom, C. F.: *Seventh Annual Report of the State Board of Health of Massachusetts,* Wright & Potter, Boston, Mass., 1876.
14. Fox, D. R., and J. C. Thayer: "Improving Reuse Site Suitability by Modifying the Soil Profile," *Proc. Third Water Reuse Symp.,* San Diego, Calif., Aug. 26–31, 1984.
15. George, M. R., G. A. Pettygrove, and W. B. Davis: "Crop Selection and Management," *Irrigation with Reclaimed Municipal Wastewater—A Guidance Manual,* California State Water Resources Control Board, July 1984.
16. Handley, L. L.: "Effluent Irrigation of California Grass," *Proc. Second Water Reuse Symp.,* Vol. 2, Am. Water Works Assoc. Research Foundation, Denver, Colo., 1981.
17. Hart, R. H.: "Crop Selection and Management," *Factors Involved in Land Application of Agricultural and Municipal Wastes,* Agr. Res. Sta., Beltsville, Md., 1974, pp. 178–200.
18. Idelovitch, E.: "Unrestricted Irrigation with Municipal Wastewater," *Proc. National Conference on Environmental Engineering,* Am. Soc. Civil Engrs., Atlanta, Ga., July 8–10, 1981.
19. Jewell, W. J., and B. L. Seabrook: *A History of Land Application as a Treatment Alternative,* EPA 430/9-79-012, Environmental Protection Agency, Office of Water Program Operations, Washington, D.C., 1979.
20. Kruzic, A. P.: "A Study of Nitrogen Removal Rates and Mechanisms in Overland Flow Wastewater Treatment," Ph.D thesis, University of California, Davis, 1984.
21. Lance, J. C., F. D. Whisler, and R. C. Rice: Maximizing Denitrification During Soil Filtration of Sewage Water," *J. Environ. Qual.,* 5:102, 1976.
22. Lau, L. S., D. R. McDonald, and I. P. Wu: "Improved Emitter and Network System Design for Reuse of Wastewater in Drip Irrigation," *Proc. Third Water Reuse Symp.,* San Diego, Calif., Aug. 26–31, 1984.
23. Loehr, R. C., and M. R. Overcash: "Land Treatment of Wastes: Concepts and General Design," *J. Environ. Eng. Div.,* ASCE, 111(2):141–160, April 1985.
24. Martel, C. J.: *Development of a Rational Design Procedure for Overland Flow Systems,* CRREL Report 82-2, Cold Regions Res. & Eng. Lab., Hanover, N.H., 1982.
25. Nolte & Associates: "Report on Land Application of Stillage Waste for Southern Ethanol Limited," Nolte & Assoc., Sacramento, Calif., March 1986.
26. Perry, L. E., E. J. Reap, and M. Gilliand: "Pilot Scale Overland Flow Treatment of High Strength Snack Food Processing Wastewaters," *Proc. Natl. Conf. Environmental Eng.,* ASCE, Environmental Eng. Div., Atlanta, July 1981, pp. 460–467.
27. Peters, R. E., C. R. Lee, and D. J. Bates: "Field Investigations of Overland Flow Treatment of Municipal Lagoon Effluent," Tech. Rep. EL-81-9, United States Army Engineers Waterways Experiment Station, Vicksburg, Miss., 1981.
28. Pound, C. E., and R. W. Crites: *Wastewater Treatment and Reuse by Land Application,* EPA 660/2-73-006b, Environmental Protection Agency, Office of Water Program Operations, Washington, D.C., 1973.
29. Reed, S. C., and R. W. Crites: *Handbook on Land Treatment Systems for Industrial and Municipal Wastes,* Noyes Data, Park Ridge, N.J., 1984.

30. Reed, S. C., R. W. Crites, and A. T. Wallace: "Problems with Rapid Infiltration—A Post Mortem Analysis," *J. Water Pollution Control Fed.*, 57(8):854–858, August 1985.
31. Seabrook, B. L.: *Land Application of Wastewater in Australia,* EPA 430/9-75-017, Environmental Protection Agency, Office of Water Program Operations, Washington, D.C., 1975.
32. Signor, D. C.: "Groundwater Sampling During Artificial Recharge: Equipment, Techniques and Data Analyses," T. Asano (ed.), *Artificial Recharge of Groundwater,* Butterworth Publishers, Stoneham, Mass., 1985, pp. 151–202.
33. Smith, R. G.: "Development of a Rational Basis for the Design and Operation of the Overland Flow Process," *Proc. National Seminar on Overland Flow Technology for Municipal Wastewater,* Dallas, Tex., Sept. 16–18, 1980.
34. Smith, R. G., and E. D. Schroeder: *Demonstration of the Overland Flow Process for the Treatment of Municipal Wastewater—Phase 2. Field Studies,* California State Water Resources Control Board, Sacramento, Calif., 1982.
35. Smith, R. G., and E. D. Schroeder: "Field Studies of the Overland Flow Process for the Treatment of Raw and Primary Treated Municipal Wastewater," *J. Water Pollution Control Fed.*, 57(7):785–794, July 1985.
36. Smith, R. G.: "Irrigation System Design," *Irrigation with Reclaimed Municipal Wastewater—A Guidance Manual,* California State Water Resources Control Board, July 1985.
37. Tchobanoglous, G.: "Report on the Stinson Beach On-site Wastewater Management District for the Period January 17, 1978 Through December 31, 1981," Stinson Beach Co., Water District, Stinson Beach, Calif., May 1982.
38. Thomas, R. E., B. Bledsoe, and K. Jackson: *Overland Flow Treatment of Raw Wastewater with Enhanced Phosphorus Removal,* EPA 600/2-76-131, Environmental Protection Agency, Office of Res. and Dev., Washington, D.C., 1976.
39. Thomas, R. E., K. Jackson, and L. Penrod: *Feasibility of Overland Flow for Treatment of Raw Domestic Wastewater,* EPA 660/2-74-087, Environmental Protection Agency, Office of Res. and Dev., Washington, D.C., 1974.
40. U.S. Dept. of the Interior: *Drainage Manual,* GPO No. 024-003-00117-1, US Govt. Printing Office, Washington, D.C., 1978.
41. U.S. Environmental Protection Agency: *Process Design Manual for Land Treatment of Municipal Wastewater,* EPA 625/1-81-013, Center for Environmental Research Information, Cincinnati, Ohio, 1981.
42. U.S. Environmental Protection Agency: *Process Design Manual for Land Treatment of Municipal Wastewater, Supplement on Rapid Infiltration and Overland Flow,* EPA 625/1-81-013a, Center for Environmental Research Information, Cincinnati, Ohio, 1984.
43. U.S. Environmental Protection Agency: *Design Manual, Onsite Wastewater Treatment and Disposal Systems,* EPA 625/1-80-012, Center for Environmental Research Information, Cincinnati, Ohio, 1980.
44. Westcot, D. W., and R. S. Ayers: "Irrigation Water Quality Criteria," *Irrigation with Reclaimed Municipal Wastewater—A Guidance Manual,* California State Water Resources Control Board, July 1984.
45. Winneberger, J. H. T.: *Septic Tank Systems, A Consultant's Tool Kit.* Vol. 1. *Subsurface Disposal of Septic Tank Effluents,* Butterworth Publishers, Stoneham, Mass., 1984.
46. Witherow, J. L., and B. E. Bledsoe: "Algae Removal by the Overland Flow Process," *J. Water Pollution Control Fed.*, 55(10):1256–1262, 1983.
47. Witherow, J. L., and B. E. Bledsoe: "Design Model for the Overland Flow Process," *J. Water Pollution Control Fed.*, 58(5):381–386, May 1986.

Chapter

8

Sludge Management and Treatment

Sludges are a common by-product from all waste treatment systems, including some of the natural processes described in previous chapters. Sludges are also produced by water treatment operations and by many industrial and commercial activities. The economics and safety of disposal or reuse options are strongly influenced by the water content of the sludge and the degree of stabilization with respect to pathogens, organic content, metals content, and other contaminants. This chapter describes several natural methods for sludge treatment and reuse. In-plant sludge processing methods such as thickening, digestion, and mechanical conditioning and dewatering methods are not included in this text: Refs. 14, 19, and 20 are recommended for that purpose.

8.1 Sludge Quantity and Characteristics

The first step in the design of a treatment or disposal process is to determine the amount of sludge that must be managed and its characteristics. A reliable estimate can be produced by deriving a solids mass balance for the treatment system under consideration. The solids input and output for every component in the system must be calculated. Typical values for solids concentrations from in-plant operations and processes are reported in Table 8.1. Detailed procedures for conducting

TABLE 8.1 Typical Solids Content from Treatment Operations[14]

Treatment operation	Typical Dry Solids %*	Typical Dry Solids $kg/10^3m^3$†§
Primary settling		
Primary only	5.00	150
Primary and waste-activated sludge	1.50	45
Primary and trickling filter sludge	5.00	150
Secondary reactors		
Activated sludge	1.25	85
Pure oxygen	2.50	130
Extended aeration	1.50	100
Trickling filters	1.50	70
Chemical plus primary sludge		
High lime (> 800 mg/L)	10.00	800
Low lime (< 500 mg/L)	4.00	300
Iron salts	7.50	600
Thickeners		
Gravity type		
Primary sludge	8.00	140
Primary and WAS‡	4.00	70
Primary and trickling filter	5.00	90
Flotation	4.00	70
Digestion		
Anaerobic		
Primary sludge	7.00	210
Primary and WAS	3.50	105
Aerobic		
Primary and WAS	2.50	80

* Percent, on a dry solids basis.
† $kg/10^3\ m^3$ = dry solids per 1000 m^3 liquid sludge.
‡ WAS = waste activated sludge.
§ See Table A.1 in appendix for conversion factors.

TABLE 8.2 Typical Composition of Wastewater Sludges[14]

Component	Untreated primary	Digested
Total solids (TS), %	5.0	10.0
Volatile solids, % of TS	65.0	40.0
pH	6.0	7.0
Alkalinity, mg/L (as $CaCO_3$)	600	3000
Cellulose, % of TS	10.0	10.0
Grease and fats (ether-soluble), % of TS	6–30	5–20
Protein, % of TS	25.0	18.0
Silica, (SiO_2), % of TS	15.0	10.0

TABLE 8.3 Nutrients and Metals in Typical Wastewater Sludges[21]

Component	Median	Mean
Total nitrogen, %	3.3	3.9
NH_4^+ (as N), %	0.09	0.65
NO_3^- (as N), %	0.01	0.05
Phosphorus, %	2.3	2.5
Potassium, %	0.3	0.4
Copper, mg/kg	850	1210
Zinc, mg/kg	1740	2790
Nickel, mg/kg	82	320
Lead, mg/kg	500	1360
Cadmium, mg/kg	16	110
PCBs, mg/kg	3.9	5.2

mass balance calculations for wastewater treatment systems can be found in Refs. 14 and 19.

The characteristics of wastewater treatment sludges are strongly dependent on the composition of the untreated wastewater and on the unit operations in the treatment process. The values reported in Tables 8.2 and 8.3 represent typical conditions only and do not provide a suitable basis for design of a specific project. The sludge characteristics must either be measured or carefully estimated from similar experience elsewhere to provide the data for final designs.

Sludges from natural treatment systems

A significant advantage for the natural wastewater treatment systems described in previous chapters is the minimal sludge production as compared with mechanical treatment processes. Any major quantities of sludge are typically the result of preliminary treatments and not of the natural process itself. The pond systems described in Chap. 4 are an exception in that, depending on the climate, sludge will accumulate at a gradual but significant rate, and its ultimate removal and disposal must be given consideration during design.

In colder climates studies have established that sludge accumulation proceeds at a faster rate, so removal may be required more than once over the design life of the pond. The results of investigations in Alaska and in Utah[26] on sludge accumulation and composition in both facultative and partial-mix aerated lagoons are reported in Tables 8.4 and 8.5.

A comparison of the values in Tables 8.4 and 8.5 with those in Tables

TABLE 8.4 Summary of Pond Sludge Accumulation Data[26]

Parameter	Facultative Ponds, Utah		Aerated Ponds, Alaska	
	A	B	C	D
Flow, m^3/day*	37,850	694	681	284
Surface, m^2	384,188	14,940	13,117	2,520
Bottom, m^2	345,000	11,200	8,100	1,500
Operated since last cleaning, years	13	9	5	8
Mean sludge depth, cm	8.9	7.6	33.5	27.7
Total solids, g/L	58.6	76.6	85.8	9.8
Volatile solids, g/L	40.5	61.5	59.5	4.8
Wastewater suspended solids, mg/L	62	69	185	170

* See Table A.1 in the Appendix for conversion factors.

8.2 and 8.3 indicates that the pond sludges are quite similar in their characteristics to untreated primary sludges. The major difference is that the content of solids, both total and volatile, is higher for most pond sludges than for primary sludge, and the fecal coliforms are significantly lower. This is reasonable in light of the very long detention time in ponds as compared with primary clarifiers. The long detention time allows for significant die-off of fecal coliforms and for some consolidation of the sludge solids. All four of the lagoons described in Tables 8.4 and 8.5 are considered to be in a cold climate. Pond systems in the southern half of the United States might expect lower accumulation rates than those indicated in Table 8.4.

TABLE 8.5 Composition of Pond Sludges[26]

Parameter*	Facultative Ponds, Utah		Aerated Ponds, Alaska	
	A	B	C	D
Total solids, %	5.9	7.7	8.6	0.89
Total solids, mg/L	586,000	766,600	85,800	9800
Volatile solids, %	69.1	80.3	69.3	48.9
Total organic carbon (TOC), mg/L	5513	6009	13,315	2651
pH	6.7	6.9	6.4	6.8
Fecal coliforms, (no. per 100 mL) $\times 10^5$	0.7	1.0	0.4	2.5
Kjeldahl nitrogen (TKN), mg/L	1028	1037	1674	336
TKN, % of total solids	1.75	1.35	1.95	3.43
Ammonia nitrogen (as N), mg/L	72.6	68.6	93.2	44.1
Ammonia nitrogen (as N), % of total solids	0.12	0.09	0.11	0.45

* See Table A.1 in the Appendix for conversion factors.

TABLE 8.6 Characteristics of Water Treatment Sludges[11]

Characteristic	Range of values
Volume, as % of water treated	< 1.0
SS concentration	0.1–1000 mg/L
Solids content	0.1–3.5%
Solids content after long-term settling	10–35%
Composition of alum sludge	
Hydrated aluminum oxide	15–40%
Other inorganic materials	70–35%
Organic materials	15–25%

Sludges from drinking water treatment

Sludges occur in drinking water treatment systems as a result of turbidity removal, softening, and filter backwash. The dry weight of sludge produced per day from softening and turbidity removal operations can be calculated with Eq. 8.1.[11]

$$S = 84.4Q(2\text{Ca} + 2.6\text{Mg} + 0.44\text{Al} + 1.9\text{Fe} + \text{SS} + A_x) \tag{8.1}$$

where S = sludge solids, kg/day
Q = design water treatment flow, m^3/s
Ca = calcium hardness removed (as $CaCO_3$), mg/L
Mg = magnesium hardness removed (as $CaCO_3$), mg/L
Al = alum dose (as 17.1% Al_2O_3), mg/L
Fe = iron salts dose (as Fe), mg/L
SS = raw water suspended solids, mg/L
A_x = additional chemicals (polymers, clay, activated carbon, etc.), mg/L

The major components of most of these sludges are the suspended solids (SS) from the raw water and the coagulant and coagulant aids used in treatment. Sludges resulting from coagulation treatment are the most common and are typically found at all municipal water treatment works. Typical characteristics of these sludges are reported in Table 8.6.

8.2 Stabilization and Dewatering

Stabilization of wastewater sludges and dewatering of almost all types of sludge are necessary for economic, environmental, and health reasons. Transport of sludge from the treatment plant to the point of disposal or reuse is a major factor in the cost of sludge management. The desirable sludge solids content for the major disposal and reuse options are presented in Table 8.7.

Sludge stabilization controls offensive odors, lessens the possibility

TABLE 8.7 Solids Content for Sludge Disposal or Reuse[21]

Disposal/reuse method	Reason to dewater	Required solids, %
Land apply	Reduce transport and other handling costs	> 3
Landfill	Regulatory requirements	> 10*
Incineration	Process requirement to reduce fuel required to evaporate water	> 26

* Greater than 20 percent in some states.

for further decomposition, and significantly reduces pathogens. Typical pathogen content in unstabilized and anaerobically digested sludge are compared in Table 3.10.

Methods for pathogen reduction

The pathogen content of sludge is especially critical when the sludge is to be used in agricultural operations or when public exposure is a concern. Two levels of stabilization are recognized by the Environmental Protection Agency, as described in Sec. 3.4.

8.3 Sludge Freezing

Freezing and then thawing a sludge will convert an undrainable jellylike mass into a granular material, which will drain immediately upon thawing. This natural process offers a cost-effective method of dewatering.

Effects of freezing

Freeze-thawing will have the same effect on any type of sludge but is particularly beneficial with chemical and biochemical sludges containing alum, which are extremely slow to drain naturally. Energy costs for artificial freeze-thawing are prohibitive, so the concept must depend on natural freezing to be cost-effective.

Process requirements

The design of a freeze dewatering system must be based on worst-case conditions to ensure successful performance at all times. If sludge freezing is to be a reliable expectation every year, the design must be based on the warmest winter during the period of concern (typically 20 years or longer). The second critical factor is the thickness of the sludge layer that will freeze within a reasonable period if freeze-thaw cycles are a normal occurrence during the winter. A common mistake in past attempts at sludge freezing has been to apply sludge in a single deep layer. In many locations a large single layer may never freeze completely to the bottom, so that only the upper portion goes through alternating freezing and thawing cycles. It is absolutely essential that the entire mass of sludge be frozen completely for the benefits to be realized, and once it has been frozen and thawed, the change is irreversible.

General equation. The freezing or thawing of a sludge layer can be described by Eq. 8.2:

$$Y = m(\Delta T \cdot t)^{1/2} \tag{8.2}$$

where Y = depth of freezing or thawing, cm (in)
m = proportionality coefficient, cm(°C · day)$^{-1/2}$
= 2.04 cm(°C · day)$^{-1/2}$
= 0.60 in(°F · day)$^{-1/2}$
$\Delta T \cdot t$ = freezing or thawing index, °C · day (°F · day)
ΔT = temperature difference between 0°C (32°F) and the average ambient air temperature during the period of interest, °C (°F)
t = time period of concern, days

Equation 8.2 has been in general use for many years to predict the depth of ice formation on ponds and streams. The proportionality coefficient m is related to the thermal conductivity, the density, and the latent heat of fusion of the material being frozen or thawed. A median value of 2.04 was experimentally determined for wastewater sludges in the range of 0 to 7 percent solids.[22] The same value is applicable to water treatment and industrial sludges in the same concentration range.

The freezing or thawing index in Eq. 8.2 is an environmental characteristic for a particular location. It can be calculated from weather records and can also be found directly in other sources.[10] The factor ΔT in Eq. 8.2 is the difference between the average air temperature during the period of concern and 0°C (32°F). Example 8.1 below illustrates the basic calculation procedure.

Example 8.1: Determination of freezing index. The average daily air temperatures for a 5-day period are listed below. Calculate the freezing index for that period.

Day	*Mean temperature, °C*
1	0
2	−6
3	−9
4	+3
5	−8

solution

1. The average air temperatue during the period is $-4°C$.
2. The freezing index for the period is:

   $\Delta T \cdot d = [0 - (-4)](5) = 20°C \cdot \text{day}$

The rate of freezing decreases with time, under steady-state temperatures since the frozen material acts as an insulating barrier between the cold ambient air and the remaining unfrozen sludge. As a result it is possible to freeze a greater total depth of sludge in a given time if the sludge is applied in thin layers.

Design sludge depth. In very cold climates with prolonged winters the thickness of the sludge layer is not critical. However in more temperate regions, particularly those that experience alternating freeze-thaw periods, the layer thickness can be very important. Calculations by Eq. 8.2 tend to converge on an 8-cm (3-in) layer as a practical value for almost all locations. At $-5°C$ (23°F) an 8-cm layer should freeze in about 3 days; at $-1°C$ (30°F) it would take about 2 weeks.

A greater depth should be feasible in colder climates. Duluth, Minnesota, for example, successfully freezes sludges from water treatment plants in 23-cm (9-in) layers.[25] It is suggested that an 8-cm depth be used for feasibility assessment and preliminary designs. A larger increment may then be justified by a detailed evaluation during final design.

Design procedures

The process design for sludge freezing must be based on the warmest winter of record to ensure reliable performance at all times. The most accurate approach is to examine the weather records for a particular location and determine how many 8-cm (3-in) layers could be frozen in each winter. The winter with the lowest total depth is then the design year. This approach might assume, for example, that the first layer is applied to the bed on 1 November each year.

Equation 8.2 is rearranged and used with the weather data to determine the number of days required to freeze the layer.

$$t = (Y/m)^2/\Delta T \tag{8.3}$$

With an 8-cm layer and $m = 2.04$, the equation becomes:

$$t = 15.38/\Delta T$$

In USCS units [3-in layer, $m = 0.6$ in $(°F \cdot \text{day})^{-1/2}$]

$$t = 25.0/\Delta T$$

Calculation methods. The mean daily air temperatures are used to calculate the ΔT value. Account is taken in the calculations of thaw periods, and a new sludge application is not made until the previous layer has frozen completely. One day is then allowed for a new sludge application and cooling, after which calculations are repeated with Eq. 8.3 to again determine the freezing time. The procedure is repeated through the end of the winter season. A tabular summary is recommended for the data and calculation results. This procedure can be easily programmed for rapid calculations with a small computer or desktop calculator.

Effect of thawing. Thawing of previously frozen layers during a warm period is not a major concern since these solids will retain their transformed characteristics. Mixing of a new deposit of sludge with thawed solids from a previously frozen layer will extend the time required to refreeze the combined layer (solve Eq. 8.3 for the combined thickness). If an extended thaw period occurs, removal of the thawed sludge cake is recommended.

Preliminary designs. A rapid method, useful for feasibility assessment and preliminary design, relates the potential depth of frozen sludge to the maximum depth of frost penetration into the soil at a particular location. The depth of frost penetration is also dependent on the freezing index for a particular location; published values can be found in the literature.[36] Equation 8.4 correlates the total depth of sludge that could be frozen if applied in 8-cm (3-in) increments with the maximum depth of frost penetration.

$$\sum Y = 1.76(F_p) - 101 \quad \text{(metric units)}$$
$$\sum Y = 1.76(F_p) - 40 \quad \text{(USCS units)} \tag{8.4}$$

where ΣY is the total depth of sludge that can be frozen in 8-cm (3-in) layers during the warmest "design" year, in centimeters (inches), and F_p is the maximum depth of frost penetration, in centimeters (inches). The maximum depth of frost penetration for selected locations in the northern United States and Canada is reported in Table 8.8.

Design limits. It can be demonstrated by use of Eq. 8.4 that sludge freezing will not be feasible unless the maximum depth of frost penetration is at least 57 cm (22 in) for a particular location. In general, that will begin to occur above the thirty-eighth parallel of latitude and will include most of the northern half of the United States with the exception of the west coast. However, sludge freezing will not be cost-

TABLE 8.8 Maximum Depth of Frost Penetration and Potential Depth of Frozen Sludge

Location	Maximum frost penetration, cm	Potential depth of frozen sludge, cm
Bangor, Maine	183	221
Concord, N.H.	152	166
Hartford, Conn.	124	117
Pittsburgh, Pa.	97	70
Chicago, Ill.	122	113
Duluth, Minn.	206	261
Minneapolis, Minn.	190	233
Montreal, Que.	203	256

effective if only one or two 8-cm (3-in) layers can be frozen in the "design" year.

A maximum frost penetration of about 100 cm (39 in) would allow sludge freezing for a total depth of 75 cm (30 in). The process should be cost-effective at that stage, depending on land and construction costs. The results of calculations by Eq. 8.4, plotted on Fig. 8.1, indicate the potential depth of sludge that could be frozen at all locations in the United States. This figure or Eq. 8.4 can be used for preliminary estimates, but final design should be based on actual weather records for the site and the calculation procedure described earlier.

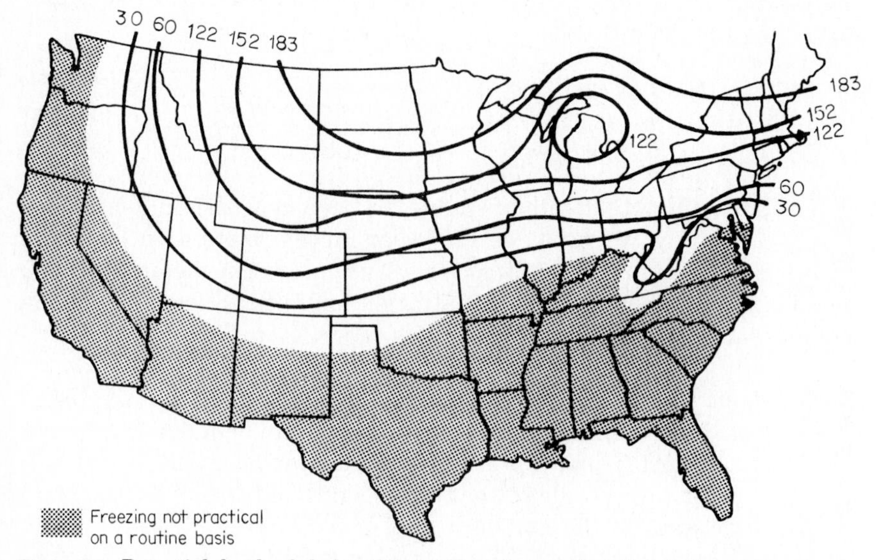

Figure 8.1 Potential depth of sludge that could be frozen if applied in 8-cm layers.

TABLE 8.9 Effects of Sludge Freezing[8,22,24,25]

Location and sludge type	Percent Solids Content	
	Before freezing	After freezing
Cincinnati, Ohio		
Wastewater sludge, with alum	0.7	18.0
Water treatment, with iron salts	7.6	36.0
Water treatment, with alum	3.3	27.0
Ontario, Canada		
Waste-activated sludge	0.6	17.0
Anaerobically digested	5.1	26.0
Aerobically digested	2.2	21.0
Hanover, N.H.		
Digested, wastewater sludge, with alum	2–7	25–35
Digested primary	3–8	30–35

Thaw period. The time required to thaw the frozen sludge can be calculated by Eq. 8.2 with use of the appropriate thawing index. Frozen sludge will drain quite rapidly. In field trials with wastewater sludges in New Hampshire, solids concentrations approached 25 percent as soon as the material was completely thawed.[22] An additional 2 weeks of drying producd a solids concentration of 54 percent. The sludge particles retain their transformed characteristics and subsequent rainfall on the bed will drain immediately, as indicated by the fact that the solids concentration was still about 40 percent 12 hours after an intense rainfall (4 cm) at the New Hampshire field trial.[22] The effects for a variety of different sludge types are reported in Table 8.9.

Sludge freezing facilities and procedures

The same basic facility can be used for water treatment sludges and wastewater sludges. The area can be designed either as a series of underdrained beds, similar in detail to conventional sand drying beds, or as deep, lined, and underdrained trenches. The Duluth, Minnesota water treatment plant uses the trench concept.[25] The sludges are pumped to the trenches on a routine basis throughout the year. Any supernatant is drawn off just prior to the onset of winter. After an initial ice layer has formed, sludge is pumped up from beneath the ice, spread in repeated layers on the ice surface, and allowed to freeze. The sand bed approach requires sludge storage elsewhere and application to the bed after the freezing season has commenced.

Effect of snow. Neither beds nor trenches require a roof or a cover. A light snowfall (less than 4 cm) would not interfere with the freezing, and the contribution of the meltwater to the total mass would be negligible. What must be avoided is application of sludge under a deep snow layer. The snow in this case will act as an insulator and retard freezing of the sludge. Any deep snow layers should be removed prior to a new sludge application.

Combined systems. If freezing is the only method used to dewater wastewater sludges, storage will be required during warm periods. A more cost-effective alternative is to combine winter freezing with polymer-assisted summer dewatering on the same bed. In a typical case winter sludge application might start in November and continue in layers until about 1 m (3 ft) of frozen material had accumulated. In most locations this would thaw and drain by early summer. Polymer-assisted dewatering could then continue on the same beds during the summer and early fall. Sludge storage in deep trenches during the warm months is better suited for water treatment operations, where putrefaction and odors are not a problem.

Sludge removal. It is recommended that the drained wastewater sludges be removed each year. Inert chemical sludges from water treatment and industrial operations could remain in place for several years. In these cases a trench 2 to 3 m (7 to 10 ft) deep could be constructed, and the dried solids residue could remain on the bottom. In addition to new construction, the sludge freezing concept can allow the use of existing conventional sand beds, which are not now used in the winter months.

Example 8.2 A community near Pittsburgh, Pennsylvania is considering freezing as the dewatering method for its estimated annual wastewater sludge production of 1500 m^3 (7 percent solids). Maximum frost penetration (from Table 8.8) is 97 cm.

solution

1. Use Eq. 8.4 to determine potential design depth of frozen sludge:

$$\sum Y = 1.76(F_p) - 101$$
$$= 1.76(97) - 101$$
$$= 70 \text{ cm}$$

2. Then determine the bed area required for freezing:

$$\text{Area} = 1500 \text{ m}^3/0.70 \text{ m} = 2143 \text{ m}^2$$

This could be provided by 16 freezing beds, each 7 m by 20 m. Allow 30 cm for freeboard, so:

$$\text{Constructed depth} = 0.70 + 0.30 = 1.0 \text{ m}$$

3. Determine the time required to thaw the 0.70-m sludge layer if average temperatures are: 10°C for March, 17°C for April, and 21°C for May.

Use Eq. 8.3 with a sludge depth of 70 cm:

$$\Delta T \cdot t = (Y/m)^2$$
$$= 1177°C \cdot \text{day}$$
$$\text{(March)} + \text{(April)} + \text{(1–17 May)}$$
$$\Delta T \cdot t = (31)(10) + (30)(17) + (17)(21) = 1177°C \cdot \text{day}$$

Therefore, the sludge layer should be completely thawed by May 18th under the assumed conditions.

Sludge quality. Although the detention time for sludge on the freezing beds may be several months, the low temperatures involved will preserve the pathogens rather than destroying them. As a result the process can only be considered as a conditioning and dewatering operation, with little additional stabilization provided. However, wastewater sludges treated in this way may be "cleaner" than sludges air-dried on typical sand beds. This is due to the rapid drainage of sludge liquid after thawing, which carries away a significant portion of the dissolved contaminants. In contrast, air-dried sludges will still contain most of the metal salts and other evaporation residues.

8.4 Reed Beds

Beds of this type using rushes, reeds, or cattails are similar in concept to the constructed wetland processes described in Chap. 6. The major difference is the use of an underdrain system in the sludge beds. Most of the operational beds have been planted with the common reed *Phragmites,* but other wetland vegetation should provide comparable performance. The use of reeds in a system of this type is claimed to be a proprietary concept by one of the proponents.[2]

The structural facility for a reed bed is similar in construction to an open, underdrained sand drying bed. Typically, a 25-cm (10-in) layer of washed gravel (20 mm) is overlain by a 10-cm (4-in) layer of filter sand (0.3 to 0.6 mm). At least 1 m (3 ft) of freeboard is provided for sludge accumulation. The selected vegetation is planted on about 30-cm (12-in) centers in the gravel layer and is allowed to become well established before the first sludge application.[1]

Function of vegetation

The root system of the vegetation absorbs water from the sludge, which is then lost to the atmosphere via evapotranspiration. The penetration of the plant stems and root system also provides a pathway for continuous drainage of water from the sludge layer. As described in Chap. 6, cattails, reeds, and similar plants are capable of transmitting oxygen from the leaf to the roots; thus there are aerobic microsites (adjacent to the roots) in an otherwise anaerobic environment, which can assist in sludge stabilization and mineralization. On the basis of work with wetland systems, it appears likely that reeds have the ability to transmit more oxygen than cattails or other species.

Design requirements

Sludge application to these reed beds is similar to the freezing process previously described in that sequential layers of sludge are applied during the operating season. The solids content of the sludge to be applied should be in the range of 3 to 4 percent, to ensure uniform distribution within the dense vegetation on the bed.[2] Solids contents lower than 3 percent are not cost-effective since the time required for drying is significantly prolonged. Using sludge at 3 to 4 percent solids would allow a 10-cm (4-in) layer of sludge to be applied every 10 days under warm, dry weather conditions.

Operating experiences

A system in New Jersey has been designed for an annual loading of 350 cm (138 in) of aerobically stabilized sludge (3 percent solids).[6] The sludge loading on a volumetric basis is 3.5 m^3 per square meter of bed area per year [86 gal/($ft^2 \cdot$ year)] or, in terms of dry solids, 105 kg/($m^2 \cdot$ year) [22 lb/($ft^2 \cdot$ year)]. Other systems, also in New Jersey, are designed for a loading rate of 1.6 m^3/($m^2 \cdot$ year) [40 gal/($ft^2 \cdot$ year)], which is in the same range as conventional sand drying beds. The final annual layer of dried sludge will be about 10 cm (4 in) deep. This residue can be left in place for several years for all chemical sludges and well-stabilized wastewater sludges; a 10-year cycle has been proposed for several systems in New Jersey.[2] At the end of the cycle the accumulated residue and the sand layer are removed, a new layer of sand is installed, and new vegetation is planted if necessary. In the typical case the vegetation will reemerge from the extensive root system in the permanent gravel layer. An annual harvest of the vegetation is recommended at the stage when the plant is dormant but the leaves have not yet been shed. The harvest will produce about 56 metric tons/ha (25 tons/acre) of vegetative material, which can be composted, burned, or otherwise disposed of.

Benefits

The major advantages of this design are the very high solids content (suitable for landfill disposal) and the low cost for sludge removal and transport on a 10-year cycle. The major disadvantage is the requirement for annual disposal of the harvested vegetation. However, over a 10-year cycle the total mass of sludge residue and vegetation requiring disposal would be less than the final sludge cake from a typical sand drying bed.

The system has been successfully operated in New Jersey with only 20 to 30 days downtime for climatic conditions. Since the dewatering benefits will be minimal during the dormant season for the plants and after harvest, it is likely that a longer out-of-service period may be necessary in areas with more severe winters than New Jersey.

Combined systems

An attractive possibility would be to combine the reed bed concept with the freezing technique described in the previous section in those locations where the latter is feasible. Such a design should not attempt to achieve the maximum possible depth of frozen sludge. A very thick layer of frozen sludge would take too long to thaw in the spring and interfere with emergence of the new vegetation. The design thickness for frozen sludge should not exceed 60 cm (24 in) in a combined system in any location. If this thickness proves inadequate for the winter sludge production, temporary storage would be necessary.

Reed bed operations

Following placement of the surface sand layer, the underdrain should be closed and the bed flooded to a depth of 10 cm (4 in). The rootstock or rhizomes for the cattails or reeds should be planted at a depth of 10 to 15 cm (4 to 6 in), and the standing water should be maintained at a depth of 10 cm for about 2 to 3 weeks to encourage plant development. The initial planting should take place in early spring, preferably in April or early May, so the first sludge application will be possible by June. It may be necessary in some locations to complete the basic construction in late summer or fall and then wait until the following spring to plant the vegetation. Harvesting can be accomplished with electric hedge clippers or similar equipment, and the plant stems should be cut so that about 8 cm (3 in) of stubble remains (these procedures were adapted from Ref. 6). Harvesting is necessary so that plant litter does not interfere with sludge distribution on the bed.

Example 8.3 A community near Pittsburgh, Pennsylvania (see Example 8.2) produces 3000 m^3 of sludge (at 3.5 percent solids) per year. Compare reed beds for dewatering with a combined reed–freezing bed system.

solution

1. Assume a 4-month freezing season: 7 months active, 5 months dormant for reeds. Design loading for reeds, $5/7 \times 2.0\ m^3/m^2 = 1.4\ m^3/m^2$; design depth for freezing, 30 cm (O.K., Example 8.3 indicates a maximum potential depth of 70

cm would be feasible). Use a 10-year cleaning cycle for the beds, so build 12 sets (10 operational, 1 for emergencies, 1 down for cleaning).

2. Calculate bed area if reed dewatering used alone:

$$\text{Area} = 3000 \text{ m}^3/1.4 \text{ m}^3/\text{m}^2$$
$$= 2143 \text{ m}^2$$

The project could use 22 active beds, each 9 × 11 m, loading 2 at a time. To allow a reserve construct 24 beds in 12 sets of pairs.

Also, storage for 5 months sludge production is needed.

$$5 \times 250 \text{ m}^3/\text{month} = 1250 \text{ m}^3$$

3. In the combined reed–freezing system, all 24 beds could be used for winter freezing at a 60-cm total depth:

$$5 \times 11 \times 0.6 \times 24 = 792 \text{ m}^3 \text{ of 3.5 percent solids sludge}$$

Thickened sludge at 7 percent could be applied for freezing, so the potential volume that could be frozen is 1584 m^3.

Use of the beds for winter freezing of thickened sludge would eliminate sludge storage requirements.

Number of units

Every system will have multiple beds, the number depending on the loading schedule and on the cleaning program. If, for example, a 10-year cycle is planned, the total number of beds constructed would be 12 to allow for one out of service each year and one available for emergencies. Current procedures suggest shutdown and a 6-month waiting period before cleaning to allow for dieoff of remaining pathogens and further stabilization. Bed shutdown should commence in early spring, the vegetation should be harvested in the fall, and the sludge residue and sand should be removed by early winter.[2]

Sludge quality

The dried material removed from these reed beds will be similar in character to composted sludge with respect to pathogen content and stabilization of organics. The long detention times combined with the final 6-month rest prior to sludge removal ensure a stable final product for reuse or disposal. In addition to its application to wastewater sludges, the reed bed concept has been used in western Europe to dewater chemical slimes and, on a pilot basis, in the United States to dewater and stabilize dredge spoil material.[2]

8.5 Vermistabilization

Vermistabilization, i.e., sludge stabilization and dewatering using earthworms, has been investigated in numerous locations and has been successfully tested on a pilot basis.[7,12] A potential cost advantage for this concept in wastewater treatment systems is provided by the capacity for stabilization and dewatering in one step, as compared with thickening, digestion, conditioning, and dewatering in a conventional process. Vermistabilization has also been successfully used with dewatered sludges and solid wastes. The concept will only be feasible for sludges that contain sufficient organic matter and nutrients to support the worm population.

Worm species

In most locations the facilities required for this procedure will be similar to an underdrained sand drying bed enclosed in a heated shelter. Studies at Cornell University evaluated four earthworm species: *Eisenia foetida, Eudrilus eugeniae, Pheretima hawayana,* and *Perionyx excavatus.* The species *E. foetida* showed the best growth and reproductive responses at temperatures in the 20 to 25°C (68 to 77°F) range. Temperatures near the upper end of that range would be necessary for optimum growth of the other species.

Worms are placed on the bed in a single initial application of about 2 kg/m^2 (0.4 lb/ft^2) live weight. Sludge loading rates of about 1 kg volatile sludge solids per square meter (0.2 lb/ft^2) per week were recommended for liquid primary and liquid waste-activated sludge.[12]

Liquid sludges used in the Cornell University tests ranged from 0.6 to 1.3 percent solids, and the final stabilized solids ranged from 14 to 24 percent total solids.[12] The final stabilized sludge had about the same characteristics regardless of the type of liquid sludge initially applied. Typical values were:

Total solids = 14 to 24 percent

Volatile solids = 460 to 550 g per kilogram total solids

Organic nitrogen = 27 to 35 g per kilogram total solids

Chemical oxygen demand = 606 to 730 g per kilogram total solids

pH = 6.6 to 7.1

Thickened and dewatered sludges have also been used in operations in Texas with essentially the same results.[7] Application of very liquid sludges (less than 1 percent solids) is feasible as long as the liquid drains rapidly so that aerobic conditions can be maintained in the unit. Final sludge removal from the unit is only required at long intervals of about 12 months.

Loading criteria

The recommended vermistabilization loading of 100 g/m^2 (0.2 lb/ft^2) per week is equivalent, for typical sludges, to a design per capita area requirement of 0.417 m^2 (4.5 ft^2). This is about 2.5 times larger than a conventional sand drying bed. The construction cost difference will be even greater since the vermistabilization bed must be covered and possibly heated. However, major cost savings are possible for the overall system, since thickening, digestion, and dewatering units may not be required if vermistabilization is used with liquid sludges.

Procedures and performance

At an operation in Lufkin, Texas thickened (3.5 to 4 percent solids content) primary and waste-activated sludge are sprayed at a rate of 0.24 kg/(m^2 · day) [0.05 lb/(ft^2 · day)] of dry solids over beds containing worms and sawdust. The latter acts as a bulking agent and absorbs some of the liquid to assist in maintaining aerobic conditions. An additional 2.5- to 5-cm (1- to 2-in) layer of sawdust is added to the bed after about 2 months. The original sawdust depth was about 20 cm (8 in) when the beds were placed in operation.

The mixture of earthworms, castings, and sawdust is removed every 6 to 12 months. A small front-end loader is driven into the bed to move the material into windrows. A food source is spread adjacent to the windrows, and within 2 days essentially all the worms have migrated to the new material. The concentrated worms are collected and used to inoculate a new bed. The castings and sawdust residue are removed, and the bed prepared is for the next cycle.[7]

Example 8.4 Determine the bed area required to utilize vermistabilization for a municipal wastewater treatment facility serving 10,000 to 15,000 people. Compare advantages of liquid versus thickened sludge.

solution

1. Assuming an activated sludge system or the equivalent, the daily sludge production would be about 1 metric ton dry solids per day. If the sludge contained about 65 percent volatile solids (see Table 8.2), the Cornell vermistabilization loading rate of 1 kg/(m^2 · week) would be equal to loading rate of 1.54 kg/(m^2 · week) of total solids. Assume 2 weeks per year downtime for bed cleaning and general maintenance. The Lufkin, Texas loading rate for thickened sludge is equal to 1.78 kg/(m^2 · week) of total solids.

2. Calculate the bed area for liquid (1 percent solids or less) and for thickened (3 to 4 percent solids) sludges.

Liquid sludge:

$$\text{Bed area} = (100 \text{ kg/day})(365 \text{ days/year})/[1.54 \text{ kg/(m}^2 \cdot \text{week)}](50 \text{ weeks})$$
$$= 4740 \text{ m}^2$$

Thickened sludge:

$$\text{Bed area} = (100 \text{ kg/day})(365 \text{ days/year})/(1.78 \text{ kg/m}^2 \cdot \text{week})(50 \text{ weeks})$$
$$= 4101 \text{ m}^2$$

3. A cost analysis would be required to identify the most cost-effective alternative. The smaller bed area for the second case is offset by the added costs required to build and operate a sludge thickener.

Sludge quality

The sludge organics pass through the gut of the worm and emerge as dry, virtually odorless castings. These are suitable for use as a soil amendment or low-order fertilizer if metal and organic chemical content are within acceptable limits (see Table 8.12 for metals criteria). There are limited quantitative data on removal of pathogens with this process. The Texas Department of Health found no *Salmonella* in either the castings or the earthworms at a vermistabilization operation in Shelbyville, Texas that received raw sludge.[7]

A market may exist for the excess earthworms harvested from the system. The major prospect is as bait for freshwater sport fishing. Use as animal or fish food in commercial operations has also been suggested. However, numerous studies[8,16] have shown that earthworms accumulate very significant quantities of cadmium, copper, and zinc from wastewater sludges and sludge-amended soils. Therefore, worms from a sludge operation should not be the major food source for animals or fish in the commercial production of food for human consumption.

8.6 Comparison of Bed-Type Operations

The physical plant for freezing systems, reed systems, and vermistabilization systems will be similar in appearance and function. In all cases a bed will be required to contain the sand or other support media; the bed must be underdrained, and a method for uniform distribution of sludge is essential. The vermistabilization beds must be covered and probably heated during the winter months in most of the United States. The other two systems require neither heat nor covers. Table 8.10 summarizes the criteria and the performance expectations for these three bed-type operations.

The annual loading rate for the vermistabilization process is much lower than the other systems discussed in this section. However, it may still be cost-effective in small to moderate-sized operations since thickening, digestion, conditioning, and dewatering can all be eliminated from the basic process design. In addition, the final product should satisfy the requirements discussed in Sec. 3.4 as a process to further reduce pathogens (PFRP). Although not specifically mentioned in the Environmental Protection Agency (EPA) regulations, the material removed from a reed bed after a 10-year cycle should also satisfy these requirements. Freezing sludge does not provide any further sta-

TABLE 8.10 Comparison of Bed-Type Operations

Factor	Freezing	Reeds*	Freezing and reeds	Worms
Sludge types	All	Nontoxic	Nontoxic	Organic, nontoxic
Bed enclosure	None	None	None	Yes
Heat required	No	No	No	Yes
Initial solids, %	1–8	3–4	3–8	1–4
Typical loading rate, $kg/(m^2 \cdot year)$†	40‡	60	50	< 20
Final solids, %	20–50§	50–90§	20–90§	15–25
Further stabilization provided	No	Some	Some	Yes
Sludge removal frequency, years	1	10¶	10¶	1

* Assumes year-round operation in a warm climate.
† Annual loading in terms of dry solids; see Table A.1 in the Appendix for conversion factors.
‡ Includes use of bed for conventional drying in summer.
§ Final solids depends on length of final drying period.
¶ The vegetation must be harvested annually.

bilization. Digestion or other stabilization of wastewater sludges is strongly recommended prior to application on freezing or reed beds to avoid odor problems.

8.7 Composting

Composting is a biological process for the concurrent stabilization and dewatering of sludges. If temperature and reaction time satisfy criteria, the final product should meet the PFRP requirements (see Sec. 3.4). There are three basic types of compost systems:

- *Windrow:* The material to be composted is placed in long rows, which are periodically turned and mixed to expose new surfaces to the air.
- *Static pile:* The material to be composted is placed in a pile and air is either blown or drawn through the pile by mechanical means. Figure 8.2 illustrates the various configurations of static pile systems.
- *Enclosed reactors:* These can range from a complete, self-contained reactor unit to structures that partially or completely enclose static pile or windrow type operations. The enclosure in these latter cases is usually for odor and climate control.

The process does not require digestion or stabilization of sludge prior to composting. Composting usually requires a dewatered sludge (less than 20 percent solids), but the process can be adjusted for variations

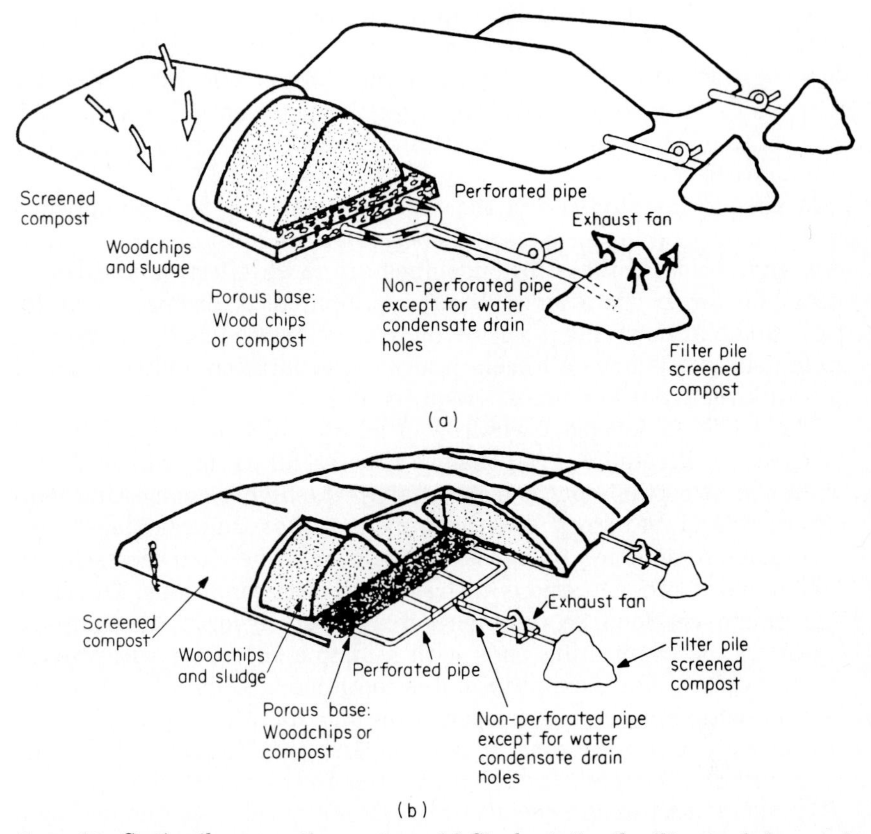

Figure 8.2 Static pile composting systems: (*a*) Single static pile; (*b*) extended aerated pile.

in initial sludge moisture content. The end product is useful as a soil conditioner (and is sold for that purpose in many locations) and has good storage characteristics.

The major process requirements include: oxygen at 10 to 15 percent, carbon/nitrogen ratio at 26:1 to 30:1, volatile solids over 30 percent, water content 50 to 60 percent, and pH at 6 to 11. High concentrations of metals, salts or toxic substances may affect the process as well as the end use of the final product. Ambient site temperatures and precipitation can have a direct influence on the operation. Most municipal sludges are too wet and too dense to be effectively composted alone, so the use of a bulking agent is necessary. Bulking agents in successful use have included wood chips, bark, leaves, corncobs, paper, straw, peanut and rice hulls, shredded tires, sawdust, dried sludge, and dried compost. Wood chips have been the most common agent and are typically separated from the finished compost mixture and used again.

The amount of bulking agent required is a function of sludge moisture content. The mixture of sludge and bulking agent should have a moisture content between 50 and 60 percent for effective composting. Sludges with 15 to 25 percent solids might require between a 2:1 and a 3:1 ratio of wood chips to sludge to attain the desired moisture content in the mixture.[32]

Mixing of the sludge and the bulking agent can be accomplished with a front-end loader for small operations. Pugmill mixers, rototillers, and special composting machines are more effective and better suited for larger operations.[33] Similar equipment is also used to build, turn, and tear down the piles or windrows. Vibratory deck, rotary, and trommel screens have all been used when separation and recovery of the bulking agent is a process requirement.

The pad area for either windrow or aerated pile composting should be paved. Concrete has been the most successful paving material. Asphalt can be suitable but it may soften at the higher composting temperatures and may itself be susceptible to composting reactions.

Outdoor composting operations have been somewhat successful in Maine and in other locations with severe winter conditions. The labor and other operational requirements are more costly for such conditions. Covering the composting pads with a simple shed roof will provide greater control and flexibility and is recommended for sites that will be exposed to subfreezing temperatures and significant precipitation. If odor control is a concern, it may be necessary to add walls to the structure and include odor control devices in the ventilation system.

The aeration piping shown in Fig. 8.2 is typically surrounded by a base of wood chips or unscreened compost about 30 to 45 cm (12 to 18 in) deep. This base ensures uniform air distribution and also absorbs excess moisture. In some cases permanent air ducts are cast into the concrete base pad. The mixture of sludge and bulking agent is then placed on the porous base material. Experience has shown that the total pile height should not exceed 4 m (13 ft) to avoid aeration problems. Typically, the height is limited by the capabilities of most front-end loaders. A blanket of screened or unscreened compost is used to cover the pile for thermal insulation and to absorb odors. About 45 cm (18 in) of unscreened or about 25 cm (10 in) of screened compost is used. Where the extended pile configuration is used, an insulating layer only 8 cm (3 in) in thickness is applied to the side that will support the next composting addition. Wood chips or other coarse materials are not recommended since the loose structure will promote heat loss and odors.

The configuration shown in Fig. 8.2 draws air into the pile and exhausts it through a filter pile of screened compost. This pile should contain about 1 m^3 (35 ft^3) of screened compost for every 3 metric tons

(3.3 tons) of sludge dry solids in the compost pile. To be effective this filter pile must remain dry; when the moisture content reaches 70 percent, the pile should be replaced.

Several systems, both experimental and operational, use positive pressure to blow air through the compost pile.[10,15] The blowers in this case are controlled by heat sensors in the pile. The advantages claimed for this approach include more rapid composting (12 versus 21 days), a higher level of volatile solids stabilization, and a drier final product. The major concern is odors, since the air is exhausted directly to the atmosphere in an outdoor operation. Positive aeration, if not carefully controlled, can result in desiccation of the lower part of the pile and therefore incomplete pathogen stabilization. The approach would seem best suited to larger operations with enclosed facilities, in which the increased control will permit realization of the potential for improved efficiency.

The time and temperature requirements for either pile or windrow composting depend on the desired level of pathogen reduction. If "significant" reduction is acceptable, then the requirement is a minimum of 5 days at 40°C (105°F) with 4 h at 55°C (130°F) or higher. If "further" reduction is necessary, then at 55°C (130°F) for 3 days for the pile method or 55°C (130°F) for 15 days with five turnings for the windrow method is required. In both cases the minimum composting time would be 21 days, and the curing time in a stockpile, after separation of the bulking agent, would be another 21 days.

A system design requires a mass balance approach to manage the input and output of solid material (sludge and bulking agent) and to account for the changes in moisture content and volatile solids. A continuing materials balance is also essential for proper operation of the system. The pad area for a composting operation can be determined with Eq. 8.5:

$$A = \frac{1.1S(R + 1)}{H} \tag{8.5}$$

where A = pad area for active compost piles, m^2 (ft^2)
S = total volume of sludge produced in 4 weeks, m^3 (ft^3)
R = ratio of bulking agent volume to sludge volume
H = height of pile, not including cover or base material, m (ft)

A design using odor control filter piles should allow an additional 10 percent of the area calculated above for that purpose. Equation 8.5 assumes a 21-day composting period but provides an additional 7 days of capacity to allow for low temperature, excessive precipitation, and malfunctions. If enclosed facilities are used and/or if positive pile aeration is planned, proportional reductions in the design area are possible.

The area calculated with Eq. 8.5 assumes that mixing of sludge and bulking material will occur directly on the composting pad. Systems designed for a sludge capacity of more than 15 dry tons/day should provide additional area for a pugmill or drum mixer.

In many locations the finished compost from the suction-type aeration will still be very moist, so spreading and additional drying are typically included. The processing area for this drying and screening to separate the bulking agent is typically equal in size to the composting area for a site in cool, humid climates. This can be reduced in more arid climates and where positive-draft aeration is used.

An area capable of accommodating 30 days of compost production is recommended as the minimum for all final curing locations. Additional storage area may be necessary depending on the end use of the compost. Winter storage may be required, for example, if the compost is only used during the growing season.

Access roads, turnaround space, and a wash rack for vehicles will all be required. If runoff from the site cannot be returned to the sewage treatment plant, then a runoff collection pond must also be included. Detention time in the pond might be 15 to 20 days with the effluent applied to the land, as described in Chap. 7. Most composting operations also have a buffer zone around the site for odor control and visual aesthetics; the size will depend on local conditions and regulatory requirements.

The aeration rate for the suction-type aerated pile is typically 14 m^3/h (8 ft^3/min) per ton of sludge dry solids. Positive-pressure aeration at higher rates is sometimes used during the latter part of the composting period to increase drying.[15] Kuter[10] used temperature-controlled positive-pressure aeration at rates ranging from 80 to 340 m^3/h (47 to 200 ft^3/min) per ton of sludge dry solids and achieved a stable compost in 17 days or less. These high aeration rates result in lower temperatures in the pile [below 45°C (113°F)]. The direction of air flow can be reversed during the latter stages to elevate the pile temperature above the required 55°C (131°F). The temperatures in the final curing pile should be high enough to ensure the required pathogen kill so the composting operation can be optimized for stabilization of volatile solids.

Monitoring is essential in any composting operation to ensure efficient operations as well as the quality of the final product. Critical parameters to be determined are:

Moisture content	In sludge and bulking material to ensure proper operations
Metals and toxics	In sludge to ensure product quality and compost reactions
Pathogens	As required by regulations

pH	In sludge, particularly if lime or similar chemicals are used
Temperature	Daily until the required number of days above 55°C (130°F) is reached, weekly thereafter
Oxygen	Initially, to set blower operation

Example 8.5 Determine the area required for a conventional extended pile composting operation for the wastewater treatment system described in Example 8.2 (1500 m^3 sludge production per year at 7 percent solids. Assume a site is available next to the treatment plant so that runoff and drainage can be returned to the treatment system.

solution

1. Use wood chips as a bulking agent. At 7 percent solids the sludge is still "wet," so a mixing ratio of at least five parts of wood chips per part of sludge will be needed. Assume top of compost at 2 m.

$$\text{4-week sludge production} = \frac{(1500)(4)}{(52)} = 115.4 \text{ m}^3$$

2. Use Eq. 8.5 to calculate the composting area:

$$A = \frac{1.1S(R + 1)}{H}$$

$$A = \frac{(1.1)(115.4)(5 + 1)}{2} = 381 \text{ m}^2$$

3. Filter piles for aeration = 10 percent of A = 0.1 381 = 38.1 m^2
4. Processing and screening area = A = 381 m^2
5. Curing area, assume 150 m^2
6. Wood chip and compost storage, assume 200 m^2
7. Roads & miscellaneous—allow 20 percent of total

$$\text{Total } A = 381 + 38.1 + 381 + 150 + 200 = 1150 \text{ m}^2$$

$$\text{Roads} = (0.2)(1150) = 230 \text{ m}^2$$

8. Total area including roads = 1380 m^2

A buffer zone might also be necessary, depending on site conditions. The area calculated here is significantly less than the area calculated in Example 8.2 for freeze drying beds. This is because composting can continue on a year-round basis but the freezing beds must be large enough to contain the entire annual sludge production.

8.8 Land Application of Sludge

The concepts described in this section are limited to those operations designed for treatment and/or reuse of the sludge via land application. Landfills and other disposal practices are covered in other texts.[17,19] Some degree of sludge stabilization is typically used prior to land application, and dewatering may be economically desirable. However, the system is designed so that the receiving land surface provides the final sludge treatment as well as utilizing the sludge organic matter and nutrients. These natural sludge management systems can be grouped in two major types, A and B.

Type A systems involve the vegetation, the soils, and the related ecosystem for treatment and utilization of the sludge. The design sludge loadings are based on the nutrient and organic needs of the site as constrained by metal, toxic substance, and pathogen contents of the sludge. Systems in this group include agricultural and forest operations, in which repetitive sludge applications are planned over a long term, and reclamation projects, in which the sludge is used to reclaim and revegetate disturbed land. The site is designed and then operated so that there are no future restrictions placed on use of the land.

Type B systems depend almost entirely on reactions in the upper soil profile for treatment. Vegetation is typically not an active treatment component, and there is no attempt to design for the beneficial utilization of sludge organic matter or nutrients. The site is often dedicated for this purpose, and there may be restrictions on the future use of the land, especially for crop production involving the human food chain. Systems receiving biodegradable sludges utilize acclimated soil organisms for that purpose and are designed for periodic loading and rest periods. Petroleum sludges and similar industrial wastes are managed in this way.

The basic feasibility of these natural sludge management options is totally dependent on the federal, state, and local regulations and guidelines that control both the sludge quality and the methodology. It is strongly recommended for all sludge management designs that the first step be determination of the possible sludge disposal/utilization options for the area under consideration. The engineer can then decide what has to be done to the sludge in the way of stabilization and dewatering so that it will be suitable for the available options. The most cost-effective combination of in-plant processes and final disposal options is not always obvious, so an iterative design procedure is required.

Concept and site selection

A preliminary evaluation should identify the available options as well as the expected physical, chemical, and biological characteristics of the sludge. The chemical characteristics will control the answers to the following questions:

1. Can the sludge be applied in a cost-effective manner?
2. Which options are technically feasible?
3. What amount of sludge can be permitted per unit area on an annual and design life basis?

4. What are the type and frequency of site monitoring and other regulatory controls imposed on the operation?

The biological characteristics of greatest concern are the presence of toxic organics and pathogens and the potential for odors during transport, storage, and application. The most important physical characteristic is the sludge moisture content. Once the amount of sludge to be managed has been estimated, it is necessary to conduct a map survey, as described in Chap. 2, to identify sites with potential feasibility for agriculture, forests, reclamation, or type B treatment. Table 8.11 presents the preliminary loading rates for the four application options. These values should only be used for this preliminary screening and *not* for design.

The land area estimates produced with the values in Table 8.11 are of the treatment area only, with no allowance for sludge storage, buffer zones, and other requirements. The preliminary screening to identify suitable sites can be a desktop analysis using commonly available information. Numerical rating procedures based on soil and groundwater conditions, slopes, existing land use, flood potential, and economic factors are described in Chap. 2 and in Refs. 17, 21, and 23. These procedures should be used to identify the most desirable sites if a choice exists. This preliminary screening is advised because it is very costly to conduct detailed field investigations on every potential site.

The final site selection is based on the technical data obtained by the site investigation, on a cost-effectiveness evaluation of capital and operating costs, and on the social acceptability of both the site and the intended sludge management option.

The requirements for pathogen reduction are discussed in Sec. 3.4. Other federal regulations include limits on cadmium (Cd) and polychlorinated biphenyls (PCBs) and set minimum soil pH levels for sludge applications where human food-chain crops are produced. Additional guidance was issued in 1982 by the EPA, the Department of Agriculture, and the Food and Drug Administration to cover the use of municipal sludge in the production of fruits and vegetables.[35]

TABLE 8.11 Preliminary Sludge Loadings for Site Identification[21]

Option	Application schedule	Typical rate, metric tons/ha
Agricultural	Annual	10
Forest	One time or at 3- to 5-year intervals	45
Reclamation	One time	100
Type B	Annual	340

Process design, type A systems

The basic design approach is based on the underlying assumption that if sludge is applied at rates that are equal to the requirements of the design vegetation over the time period of concern, there should be no greater impact on the groundwater than from normal agricultural operations. The design loading, initially based on nutrient requirements, is modified as required to satisfy limits on metals and toxic organics. As a result of this design approach extensive monitoring should not be required and the use of sludge by private farmers is possible. As the loading increases, as it may in forests and on dedicated sites, the potential for nitrate contamination of the groundwater increases, and it is then usually necessary to design a municipally owned and operated site to ensure proper management and monitoring.

Metals. Most regulations limit the total amount of lead, zinc, copper, nickel, and cadmium that can be applied to crop land. In general, these limits are based on the cation exchange capacity (CEC) of the soil as an indicator of the relationship between total metals addition and the soil's ability to control metal uptake by plants. The CEC of a soil is a measure of the net negative charge associated with both clay minerals and organic matter. However, this does not mean that the metals are retained by the exchange complex as a cation. Very few of the metals in sludge are, in fact, present as exchangeable cations. The CEC was still chosen as an indicator since it is easily measured and is related to those soil properties that do minimize plant availability of sludge metals in a soil. Table 8.12 presents the recommended CEC limits for the major metals of concern. The CEC categories of less than 5, 5 to 15, and more than 15 meq per 100 g correspond to sands, sandy loams, and silt loams, but regional differences in this pattern can occur.

Most crops do not accumulate lead, but there is concern for the direct

TABLE 8.12 Cumulative Limits for Metals Applied to Agricultural Crop Land[21]

	Metal Loading, kg/ha				
Soil type	Pb	Zn	Cu	Ni	Cd
Sand < 5 meq per 100 g	560	280	140	140	4.9
Sandy loam 5–15 meq per 100 g	1121	560	280	280	10.0
Silt loam > 15 meq per 100 g	2242	1121	560	560	20.0

NOTE: Interpolation should be used to obtain values in the 5–15 CEC range, assuming soil will be maintained at a pH of 6.5 or greater.

ingestion of lead by grazing animals. Most crops will accumulate zinc, copper, and nickel, but the levels in Table 8.12 are intended to prevent phytotoxic levels. These three metals will tend to be toxic to a crop before their concentration in the plant matter becomes a threat to animal or human health.

The federal criteria for cadmium limit the annual amount that can be applied in addition to the cumulative value in Table 8.12. These values for different crop categories are:[21]

Tobacco, leafy vegetables, root crops	0.50 kg/(ha · year) [0.45 lb/(acre · year)]
Other human foods	0.56 kg/(ha · year) [0.5 lb/(acre · year)]
Animal feed	No annual limit

These cadmium limits are based on worst case assumptions (vegetarians growing 100 percent of their food on a sludge-treated acid soil). It is necessary with cadmium to impose some extra safety factor, since this metal may accumulate to levels that may pose risks to humans without causing phytotoxicity in the plant. Some states have adopted more stringent metal limits than those presented above so it is essential to consult local regulations prior to design of a specific system.

The design loading for the metal of concern on an annual or cumulative basis can be calculated by Eq. 8.6:

$$R_m = K_m L_m / C_m \tag{8.6}$$

where R_m = metal-limited sludge application rate for the time interval selected, metric tons/ha (lb/acre)

K_m = 0.001 (metric units) [0.002 (USCS units)]

L_m = metal limit of concern, annual or lifetime for cadmium lifetime for other metals, kg/ha (lb/acre)

C_m = metal content of sludge, as a decimal fraction (e.g., for sludge with 50 ppm Cd, $C_m = 0.00005$)

Phosphorus. Some states require that the nutrient-limited sludge loading be based on the phosphorus needs of the design vegetation to ensure even more positive groundwater protection. This provides a safety factor against nitrate contamination since most sludges contain far less phosphorus than nitrogen but most crops require far more nitrogen than phosphorus, as shown in Table 7.5. If optimum crop production is a project goal, this approach will require supplemental nitrogen fertilization. Equation 8.7 can be used to determine the phosphorus-limited sludge loading. It is based on the common assumption[21] that only 50 percent of the total phosphorus in the sludge is available.

$$R_p = K_p U_p / C_p \tag{8.7}$$

where R_p = phosphorus-limited annual sludge application rate, assuming 50% availability in the sludge, metric ton/ha (ton/acre)
K_p = 0.002 (metric units) [0.001 (USCS units)]
U_p = annual crop uptake of phosphorus [see Table 7.5 for selected values, Chap. 3 of this text for further discussion, and Ref. 21 for more exact data for midwestern crops kg/ha (lb/acre)]
C_p = total phosphorus in sludge, as a decimal fraction (equation has already been adjusted for 50% availability)

Nitrogen. Calculation of the nitrogen-limited sludge loading rate is the most complicated of the calculations involved because of the various forms of nitrogen available in the sludge, the various application techniques, and the pathways nitrogen can take following land application. Most of the nitrogen in municipal sludges will be in the organic form, tied up as protein in the solid matter. The balance of the nitrogen will be in the ammonia (NH_3) form. When liquid sludges are applied to the soil surface and allowed to dry before incorporation, about 50 percent of the ammonia content will be lost to the atmosphere through volatilization.[28] As a result, only 50 percent of the ammonia is assumed to be available for plant uptake if the sludge is surface-applied. If the liquid sludge is injected or immediately incorporated, 100 percent of the ammonia is considered to be available.

The availability of the organic nitrogen is dependent on the "mineralization" of the organic content of the sludge. Only a portion of the organic nitrogen will be available in the year the sludge is applied, and a decreasing amount will continue to be available for many years thereafter. The rate will be higher, the higher the initial organic nitrogen content. The rate drops rapidly with time so for almost all sludges it is down to about 3 percent per year of the remaining organic nitrogen, after the third year.

Within the first few years of a sludge application the nitrogen contribution from this source can be very significant. It is essential to include this factor when the design is based on annual applications and nitrogen is the potential limiting parameter. The nitrogen in the sludge that is available (to plants) during the application year is given by Eq. 8.8, and the available nitrogen from that same sludge in subsequent years is given by Eq. 8.9. When annual applications are planned, it is necessary to repeat the calculations using Eq. 8.9 and then add the results to those of Eq. 8.8 to determine the total available nitrogen in a given year. These results will converge on a relatively constant value after 5 to 6 years if sludge characteristics and application rates remain about the same.

Available nitrogen in application year is given by:

$$N_a = K_N[NO_3 + k_v(NH_4) + f_n(N_0)] \tag{8.8}$$

where N_a = plant-available nitrogen in the sludge during the application year, kg/metric ton dry solids (lb/ton dry solids)
K_N = 301 (metric units)
= 602 (USCS units)
NO_3 = fraction of nitrate in the sludge (as a decimal)
k_v = volatilization factor
= 0.5 for surface-applied liquid sludge
= 1.0 for incorporated liquid sludge and dewatered digested sludge applied in any manner
NH_4 = fraction of ammonia nitrogen in sludge (as a decimal)
f_n = mineralization factor for organic nitrogen in first year $n = 1$ (see Table 8.13 for values)
N_0 = fraction of organic nitrogen in sludge (as a decimal)

Nitrogen available in subsequent years is given by

$$N_{pn} = K_N[f_2(N_0)_2 + f_3(N_0)_3 + \ldots f_n(N_n)] \quad (8.9)$$

where N_{pn} = plant-available nitrogen available in year n from mineralization of sludge applied in a previous year, kg/metric ton (lb/ton) dry solids
$(N_o)_n$ = decimal fraction of the organic nitrogen remaining in the sludge in year n

The annual nitrogen-limited sludge loading is then calculated by Eq. 8.10:

$$R_N = U_N/(N_a + N_{pn}) \quad (8.10)$$

where R_N = annual sludge loading in year of concern, metric tons/ha (tons/acre)

TABLE 8.13 Mineralization Rates for Organic Nitrogen in Wastewater Sludges[21,28]

Time after sludge application, years	Mineralization rates, %		
	Raw sludge	Anaerobic digested	Composted
1	40	20	10
2	20	10	5
3	10	5	3
4	5	3	3
5	3	3	3
6	3	3	3
7	3	3	3
8	3	3	3
9	3	3	3
10	3	3	3

U_N = annual crop uptake of nitrogen (see Tables 7.5 and 7.6), kg/ha (lb/acre)
N_a = plant-available nitrogen from current year's sludge, from Eq. 8.8, kg/metric ton (lb/ton) of dry solids
N_{pn} = plant-available nitrogen from mineralization of all previous applications, kg/metric ton (lb/ton) of dry solids

Land area calculation. Equations 8.6, 8.7, and 8.10 should be solved to determine the parameter limiting the sludge loading. Some regulatory authorities require limits on constituents other than nitrogen, phosphorus, or metals. In such a case it will be necessary to determine the allowable sludge loading using a variation of either Eq. 8.6 or Eq. 8.7, depending on whether a concentration limit or a removal mechanism is specified. The limiting parameter for design will then be the constituent that results in the lowest calculated sludge loading. The application area can then be determined with Eq. 8.11. The area calculated with this equation is only the actual application area; it does not include any allowances for roads, buffer zones, and seasonal storage.

$$A = Q_s/R_L \tag{8.11}$$

where A = application area required, ha (acre)
Q_s = total sludge production for the time period of concern, metric tons (tons) of dry solids
R_L = the limiting sludge loading rate as defined by previous equations, metric tons/(ha · year) [ton/(acre · year)] for annual systems or for the time period of concern

It is not likely that the design procedure described above will result in the ideal balance of nitrogen, phosphorus, and potassium for optimum crop production in an agricultural operation. The amounts of these nutrients in the sludge to be applied should be compared with the fertilizer recommendations for the desired crop yield, and supplemental fertilizer should be applied if necessary. Reference 21 gives typical nutrient requirements for crops in the midwestern states; agricultural agents and extension services can provide similar data for most other locations.

Annual applications are a common practice on agricultural operations. Forested systems typically apply sludge at 3- to 5-year intervals owing to the more difficult site access and distribution. The total sludge loading is designed with the equations presented above. However, because of mineralization of the larger single application, there may be a brief period of nitrate loss during the year of the sludge application. Since trees are not a food chain crop, the metal limits presented earlier do not apply if topography and soils on the site would preclude any potential for future agricultural use. For similar reasons, maintenance of the soil pH at 6.5 is not essential, and regulatory authorities should be asked for a waiver. Reference 21 suggests the following metal limits for forest system designs:

Metal	Maximum lifetime loading	
	kg/ha	lb/acre
Pb	1120	1000
Zn	560	500
Cu	280	250
Ni	280	250

Cadmium is not included as long as the site has no potential for future agriculture. If the site has such potential, the EPA limits presented earlier should be used.

Reclamation and revegetation of disturbed land generally require a large quantity of organic matter and nutrients at the start of the effort in order to be effective. As a result, the sludge application is typically designed as a single, one-time operation, and the lifetime metal limits given in Table 8.12 are controlling on the assumption that the site might some day be used for agriculture. A single large application of sludge may result in a temporary nitrate impact on the site groundwater. That impact should be brief and is preferable to the long-term environmental impacts from the unreclaimed area.

When cadmium controls the sludge loading, the same total application area will be needed for either agricultural or reclamation projects. Forested systems may realize higher loadings and/or smaller treatment areas if the regulatory authorities waive the cadmium limits.

Forest systems may require the largest total land area of the three concepts because of access and application difficulties. Application of liquid sludge has been limited to tank trucks with sprinklers or spray guns. The maximum range of these devices is about 37 m (120 ft). To ensure uniform coverage the site would need a road grid on about 76-m (250-ft) centers or would have to limit applications to 37 m (120 ft) on each side of the existing road and firebreak network.

Experience has shown that tree seedlings do poorly in fresh sludge.[13] It may be necessary to wait for 6 months before planting to allow for aging of the sludge. Weeds and other undergrowth will crowd out new seedlings so herbicides and cultivation may be necessary for at least 3 years.[29] Sludge spraying on young deciduous trees should be limited to their dormant period to avoid heavy sludge deposits on the leaves.

Example 8.6 Find the area required for sludge application in an agricultural operation. Assume the following characteristics and conditions: anaerobically digested sludge production, metric tons/day of dry solids; sludge solids content 7 percent; total nitrogen 3 percent; ammonium 2 percent; nitrate 0; lead 430 ppm; cadmium 18 ppm; nickel 80 ppm; copper 400 ppm; zinc 900 ppm. The sandy site soils have a CEC of 5 meq per 100 g. A marketable crop is not intended, but the site will be planted with a grass mixture. It is expected that orchard grass will eventually dominate. The local regulatory authorities accept the EPA metal limitations and allow a design based on nitrogen fertilization requirements. A parcel of land is available within 6 km of the treatment plant.

solution

1. A preliminary cost analysis indicates that transport of the liquid sludge to the nearby site will be cost-effective, so further dewatering will not be required and surface application will be used.

2. For a CEC of 5 meq per 100 g the metal limits are (from Table 8.12): lead 560 kg/ha, zinc 280 kg/ha, copper 140 kg/ha, nickel 140 kg/ha, cadmium 4.9 kg/ha. The annual nitrogen uptake of the grass would be (from Table 7.5) 224 kg/(ha · year). Since the grass will not be used in the human food chain, the annual cadmium limits would not apply. The mineralization rates (from Table 8.13) for anaerobically digested sludge would be: 20, 10, 5, and 3 percent, etc.

3. The lifetime metal loadings are calculated with Eq. 8.6:

$$R_m = 0.001\ L_m/C_m$$

$$R_m = (0.001)(560\ \text{kg/ha})/(0.000430)$$

$$= 1302 \text{ metric tons dry sludge per hectare}$$

Similarly:

Zn: $R_m = 311$ metric tons/ha

Cu: $R_m = 350$ metric tons/ha

Ni: $R_m = 1750$ metric tons/ha

Cd: $R_m = 272$ metric tons/ha

Cadmium results in the lowest loading and is therefore the limiting metal parameter. As a result, only 272 metric tons/ha of sludge can be applied during the useful life of the site if sludge conditions remain the same.

4. Use Eqs. 8.8 and 8.9 to calculate the available nitrogen in the sludge. Since the liquid sludge will be surface-applied, there will be volatilization losses, and k_v will equal 0.5. Assume that organic nitrogen equals total nitrogen less ammonium nitrogen.

$$N_a = K_N[NO_3 + k_v(NH_4) + f_n(N_0)]$$

$$N_a = 301[0 + (0.5)(0.02) + (0.20)(0.01)]$$

$$= 301[0.012]$$

$$= 3.6 \text{ kg/per metric ton of day solids}$$

The residual nitrogen in this sludge in the second year is:

$$(N_0)_1 - f_1 N_0 = (0.01) - (0.20)(0.01)$$

$$= 0.008 \text{ (as a decimal fraction)}$$

the second-year mineralization would be:

$$f_2(N_0)_2 = (0.10)(0.008) = 0.0008$$

residual nitrogen in third year:

$$(N_0)_3 = (0.008) - (0.0008) = 0.0072$$

Similarly:

Mineralization in third year = 0.0004
fourth year = 0.0002
fifth year = 0.0002, etc.

The total available nitrogen in the second year is the second-year contribution plus the residual from the first year.

$$(N_a)_2 = (N_a)_1 + (K_N N_0)_2$$

$$(N_a)_2 = 3.6 + (301)(0.008) = 6.0 \text{ kg/per metric ton of dry sludge}$$

Similarly:

$$(N_a)_3 = (N_a)_1 = K_N[(N_0)_2 = (N_0)_3]$$

$$(N_a)_3 = 3.6 + (301)(0.008 + 0.0072) = 8.2 \text{ kg/metric ton}$$

$$(N_a)_4 = 8.3 \text{ kg/metric ton}$$

$$(N_a)_5 = 8.4 \text{ kg/metric ton, etc.}$$

Assuming the sludge characteristics stay the same, the available nitrogen will remain at about 8.4 kg/per metric ton of dry sludge from the fifth year on.

5. Use Eq. 8.10 to calculate the annual nitrogen-limited sludge loading. Use 8.4 kg/metric ton as the steady-state value from step 4 above.

$$R_N = \frac{U_N}{N_a + N_{pn}}$$

$$R_N = 224/8.4$$

$$= 27 \text{ metric tons/(ha} \cdot \text{year) of dry sludge}$$

It would be possible to apply higher loadings during the first 2 years if desired, since the full cumulative effects of mineralization are not realized until the third year.

6. Use Eq. 8.11 to find the required application area. Since food chain crops are not involved, the annual loading is based on the nitrogen limits.

$$A = \frac{Q_s}{R_L}$$

$$A = (3 \text{ metric tons/day})(365 \text{ days/year})/[27 \text{ metric tons/(ha} \cdot \text{year)}]$$

$$= 40.5 \text{ ha}$$

7. Determine the useful life of the site for sludge application. This will ensure that there are no restrictions on potential future land uses, including production of human food crops. The cadmium-limited sludge loading calculated in step 3 will control:

$$\text{Useful life} = 272 \text{ metric tons/ha}/27 \text{ metric tons/year}$$

$$= 10 \text{ years}$$

A system design for a reclamation site would typically use a single sludge application. The total annual sludge production is 1095 metric tons/year (see step 5 above). At a single loading of 272 metric tons · ha there would be a requirement for 4 ha of land requiring reclamation each year. Reclamation project designs must ensure that sufficient land will be available for each year of the intended operational life.

A forested site might not be constrained by cadmium at all. Using the criteria presented earlier in the text, zinc would be the limiting metal and would allow a total sludge loading of 622 metric tons/ha. The useful life at 27 metric tons/(ha · year) would be 23 years. If the regulatory authorities approved, the sludge might be applied at 100 metric tons/ha every 4 years.

Design of type B systems

The design of type B systems typically includes all the factors discussed for type A systems, since metals and nutrients may still control the sludge loading and the useful life of the site. In addition, the sludges intended for type B systems may contain a larger fraction of biodegradable material than typical municipal sludges, and they have significant concentrations of toxic or hazardous substances. These materials, more common in petroleum and many industrial sludges, are quite often organic compounds. If they are degradable, their presence may control the frequency as well as the size of the design unit loading on the system. If they are nondegradable, the application site should more properly be considered as a disposal or containment operation; information on such systems can be found elsewhere.[27,31]

The primary mechanism for degradation of organic chemicals in soil results from the activity of soil microorganisms. Volatilization may be significant for some compounds,[3,9] and plant uptake may be a factor if vegetation is a system component, but biological reactions are the major treatment mechanism.

Design approach. The design approach for these organic materials is based on their half-life in the soil system. This is analogous in some respects to the mineralization rate approach for nitrogen management. If, for example, a substance in the sludge had a 1-year half-life and the sludge were applied on an annual basis, half of the mass of the substance would still be left in the soil at the end of the first year. At the end of the second year three-quarters of the annual mass applied would still be in the soil, and so forth, until at the seventh year the mass remaining in the soil would be very close to the amount of the annual application.

It is suggested for those compounds with a half-life of up to 1 year, that the amount allowed to accumulate in the soil should not exceed twice the amount applied annually.[3,4] This can be achieved in all cases by adopting an application schedule that is equal to one half-life of the substance of concern.

The biological reactions in the soil are influenced by soil texture and structure, moisture content, temperature, oxygen level, nutrient status, pH, and type and number of microorganisms present. The optimum conditions for all these factors are essentially the same as those required for successful operation of an agricultural type A system. An aerobic soil with a pH of 6 to 7, a temperature of at least 10°C (50°F), and soil moisture at field capacity would represent near-optimum conditions for most situations. An additional special concern with toxic organics is their impact on the soil microbes. A unit loading that is too high may actually sterilize the soil. Mixing of the soil and the sludge reduces this risk and promotes aeration and contact between the microbes and the waste. As a result of this need for mixing, surface vegetation is not typically a treatment component in systems designed for short-half-life sludges.

Data requirements. Characterization of the sludge constituents is a critical first step in design, especially if potentially toxic or hazardous organic compounds are present. Essential data include: inorganic chemicals, electrical conductivity, pH, titratable acids and bases, moisture (water) content, total organic matter, volatile organic compounds, extractable organic compounds, residual solids, and a biological assessment to determine acute and genetic toxicity. The inorganic chemicals would include the same metals, nutrients, and halides and other salts that would be included in an analysis for type A designs.

Half-life determination. The degradation and half-life of complex organic compounds are typically determined in the laboratory by a series of soil respirometer tests. Representative samples of soil and sludge are mixed in a proportional range and placed in sealed flasks, which in turn are placed in an incubation chamber. Humidified, carbon dioxide–free air is passed through each flask. The carbon dioxide evolved from microbial activity in the flask is picked up by the air and then collected in columns containing 0.1*N* sodium hydroxide. The sodium hydroxide solutions are changed about three times a week and then titrated with hydrochloric acid. Detailed procedures can be found in Refs. 3 and 30. The typical incubation period is up to 6 months. The control tests are run at 20°C (68°F), but if field temperatures are expected to vary by more than 10°C, the half-life at these other temperatures should also be determined. In some cases it is desirable to verify laboratory results with pilot studies in the field. Soil samples are taken on a routine basis after application and mixing of the sludge and soil. The analysis should include total organics as well as compounds of specific concern.

In addition to measurements of carbon dioxide evolution by the respirometer tests, it is recommended that the organic fractions of the original sample and that of the final soil-sludge mixture be determined. The degradation rates are then determined with Eqs. 8.12 and 8.13 below.

For total carbon degradation:

$$D_t = (0.27)[CO_2]_w - [CO_2]_s)/C \tag{8.12}$$

where D_t = fraction of total carbon degraded over time t
$[CO_2]_w$ = cumulative CO_2 evolved by soil-waste mixture
$[CO_2]_s$ = cumulative CO_2 evolved by unamended soil
C = carbon applied with the sludge

For organic carbon degradation:

$$D_{t,o} = 1 - (C_{r,o} - C_s)/C_{a,o} \tag{8.13}$$

where $D_{t,o}$ = fraction of organic carbon degraded over time t
$C_{r,o}$ = amount of residual carbon in the organic fraction of the final sludge-soil mixture
C_s = amount of organic carbon extracted from the unamended soil
$C_{a,o}$ = amount of carbon in the organic fraction of the applied sludge

The degradation rate of individual organic subfractions are also determined by Eq. 8.13. The half-life for the total organics or for a specific waste is determined by Eq. 8.14.

$$t_{1/2} = 0.5\ t/D_t \tag{8.14}$$

where $t_{1/2}$ = half-life of the organics of concern, days
t = time period used to produce the data for Eqs. 8.12 or 8.13, days
D_t = fraction of carbon degraded over time t

If vegetation is to be a routine treatment component in the operational system, greenhouse and/or pilot field studies are necessary to evaluate toxicity and develop optimum loading rates. Greenhouse studies are easier and less costly to run, but field studies are more reliable. Systems designed only for soil treatment need not be concerned unless vegetation is planned as a postclosure activity.

Since a range of sludge-soil mixtures is tested in the respirometers, it is also possible to determine the concentration at which acceptable microbial activity occurs. It is then possible to determine the annual loading from this value and the previously determined half-life:

$$C_{yr} = 0.5\ C_c/t_{1/2} \tag{8.15}$$

where C_{yr} = annual application rate for the organic of concern, kg/(ha · year) [lb/(acre · year)]
C_c = critical concentration at which acceptable microbial toxicity occurs, kg/ha (lb/acre)
$t_{1/2}$ = half-life of the organic of concern, years

The loading rate is then calculated by a variation of Eq. 8.6:

$$R_{oc} = C_{yr}/C_w \tag{8.16}$$

where $R_{o,c}$ = loading rate limited by organics, kg/(ha · year) [lb/(acre · year)]
C_{yr} = annual application rate for organic of concern (Eq. 8.15), kg/(ha · year) [lb/(acre · year)]
C_w = fraction of the organic of concern in the sludge (as a decimal)

If the half-life of the organic of concern is less than 1 year, the $R_{o,c}$ calculated with Eq. 8.15 may be applied on a more frequent schedule. In this case the number of applications becomes:

$$N = 1/t_{1/2} \tag{8.17}$$

where N is the number of applications per year and $t_{1/2}$ is half-life, in years.

The land area required is then determined by Eq. 8.11. As with type A systems, the calculations are performed for nutrients, metals, and other potentially limiting factors. The limiting parameter for design is then the constituent requiring the largest land area as calculated by Eq. 8.11.

Loading nomenclature. Depending on industrial conventions and practices, the loading rates and application rates used in the design calculations may be expressed in a variety of units. For example, in the petroleum industry it is common to express the loading in terms of barrels per hectare. In most cases the sludge is mixed with the surface soil. This surface zone, termed the *incorporation zone,* is typically 15 cm (6 in) thick. As a result the loading is also often expressed as kilograms per meter of incorporation zone or as a percentage of a contaminant (on a mass basis) in the incorporation zone. The calculations below illustrate the various possibilities.

One barrel (bbl) of oil contains 159 L, which is about 143 kg of oil. One cubic meter of "typical" soil contains about 1270 kg of soil (1 ft^3 = 80 lb of soil).

One hectare of treatment area with a 15-cm (6-in) incorporation zone contains: (0.15)(10,000) = 1500 m^3/ha (21, 437 ft^3/acre). At 100 barrels of oil per hectare (40 bbl/acre), the mass loading would be (100)(143)/1500 = 9.53 kg/m^3 (0.6 lb/ft^3) of incorporation zone. At 500 bbl/ha, (200 bbl/acre), the mass loading (percent basis) would be: (500)(143)/(1500)(1270) = 3.75 percent oil in the incorporation zone.

Example 8.7 Find the land area required for treatment of a petroleum sludge produced at a rate of 5 metric tons/day containing 15 percent critical organics.

The following data were obtained with respirometer tests:

$$\text{Applied carbon (C)} = 3000 \text{ mg}$$
$$CO_2 \text{ produced (90 days)} = 1500 \text{ mg (waste + soil)}$$
$$= 100 \text{ mg (soil only)}$$

A field test indicated the critical application C_c for maintenance of the soil microbes was 71,500 kg/(ha · year) (3.75 percent).

solution

1. Use Eq. 8.12 to determine evolved carbon dioxide on a total carbon basis:

$$D_t = 0.27\,[(CO_2)_w - (CO_2)_s]/C$$
$$D_{90} = (0.27)(1500 - 100)/3000$$
$$= 0.13$$

2. Determine the half-life for the organic compounds using Eq. 8.14:

$$t_{1/2} = 0.5\, t/D_t$$
$$t_{1/2} = (0.5)(90)/0.13$$
$$= 346 \text{ days}$$
$$= 0.95 \text{ year}$$

3. Determine the application rate for the critical compounds using Eq. 8.15:

$$C_{yr} = 0.5\, C_c/t_{1/2}$$
$$C_{yr} = 0.5\,(71{,}500)/.95$$
$$= 37{,}632 \text{ kg/(ha} \cdot \text{year)}$$

4. Determine the organic-controlled loading rate with Eq. 8.16:

$$R_{oc} = C_{yr}/C_w$$
$$R_{oc} = 37{,}632/0.15$$
$$= 250{,}880 \text{ kg/(ha} \cdot \text{year)}$$
$$= 251 \text{ metric tons/(ha} \cdot \text{year)}$$

5. Determine the land area required by Eq. 8.11:

$$A = Q_s/R_L$$
$$A = (5)(365)/251$$
$$= 7.3 \text{ ha}$$

6. To complete the design calculations, the area required for nutrients, metals, and any other limiting substances should be determined. The largest of these calculated areas will then be the design treatment area.

Site details for type B systems. The site selection procedure and design for a type B system will depend on whether the site is to be permanently dedicated for a treatment/disposal operation or if it is to be restored and made available for unrestricted use following its operational life. A system of the former type may be operated as a treatment system, but ultimately one or more of the waste constituents will exceed the specified cumulative limits, so the site must be planned as a disposal operation. Criteria for these disposal operations can be found in Refs. 27 and 31.

The general site characteristics are similar both for type A and type B systems. A major difference is often the control of runoff. Off-site runoff is not generally permitted for either type of operation; however in the case of agricultural sludge operations, runoff is contained but then may be allowed to infiltrate the application site. Runoff is a more serious concern for type B operations since the sludge may contain mobile toxic or hazardous constituents.

The site is typically selected, or constructed, on a gentle slope (1 to 3 percent) and subdivided into diked plots. The purpose is to induce controlled runoff and ensure minimum infiltration and percolation. A complete hydrographic analysis is required to determine the criteria for design of collection channels, retention basins, and structures to prevent off-site runoff from entering the site. Such designs should be based on the peak discharge from a 25-year storm, and the retention basins should be designed for a 25-year, 24-hour return period storm. The discharge pathway from the retention basin will depend on the composition of the water. In many cases it may be land applied using one or more of the techniques described in Chap. 7. Special treatments may be required for critical materials; sprinklers or aeration in the retention basin are often used to reduce the concentration of volatile organics.

If clay or other liners are a site requirement, then underdrains will be necessary. Underdrains may also be required to control groundwater levels in an unlined site and to ensure maintenance of aerobic conditions in the incorporation zone. Any water collected with these drains must also be retained and possibly treated.

The site design must also consider the application method to be used and appropriate access for vehicles must be provided. Sprinklers and portable spray guns have been used with liquid sludges. In this case the civil engineering aspects of site design are quite similar to those for the overland flow concept described in Chap. 7. Dry sludges can be spread and mixed with the same type of equipment that would be used for type A operations.

On-site temporary storage may also be a requirement, particularly in colder climates. Optimal soil temperatures for microbial activity are 20°C (68°F) or higher. If lower temperatures are expected, the interval between applications can be extended (as determined by field or respirometer tests), or the sludge can be stored during the cold periods.

The soil temperatures for bare soil surfaces are commonly higher than the ambient air temperature by 3 to 5°C during daylight hours. Surface soils at many land treatment sites may exceed ambient temperatures by 5 to 10°C owing to microbial activity and increased radiation absorption when dark, oily wastes are incorporated.[13] In the general case it can be assumed that active degradation is possible when the ambient air temperatures are 10°C (50°F) or higher and no frost remains in the soil profile. On this basis the operational season for a type B system may be slightly longer than for an agricultural type A system in the same location.

REFERENCES

1. Banks, L., and S. F. Davis: "Desiccation and Treatment of Sewage Sludge and Chemical Slimes with the Aid of Higher Plants," *Proc. Symp. Municipal and Industrial Sludge Utilization and Disposal,* Rutgers University, Atlantic City, N.J., Apr. 6–8, 1983, pp. 172–173.
2. Banks, L., and S. F. Davis: "Wastewater and Sludge Treatment by Rooted Aquatic Plants in Sand and Gravel Basins," *Proc. Workshop on Low Cost Wastewater Treatment,* Clemson University, Clemson, S.C., pp. 205–218, April 1983.
3. Brown, K. W.: *Hazardous Waste Land Treatment,* Environmental Protection Agency, Rep. SW-874, EPA Office of Solid Waste and Emergency Response, Washington, D.C., April 1983.
4. Burnside, O. V.: "Prevention and Detoxification of Pesticide Residues in Soils," *Pesticides in Soil and Water,* Soil Scientists of America Inc., Madison, Wis., 1974, pp. 387–412.
5. Cole, D. W., C. L. Henry, P. Schiess, and R. J. Zasoski: "The Role of Forests in Sludge and Wastewater Utilization Programs," *Proc. 1983 Workshop on Utilization of Municipal Wastewater and Sludge on Land,* University of California, Riverside, 1983, pp. 125–143.
6. Costic & Associates: *Engineers Report—Washington Township Utilities Authority Sludge Treatment Facility,* Costic & Associates, Long Valley, N.J., 1983.
7. Donovan, J.: *Engineering Assessment of Vermicomposting Municipal Wastewater Sludges,* Environmental Protection Agency EPA-600/2-81-075 (available as PB81-196933 from National Technical Information Service, Springfield, Va.), June 1981.
8. Farrell, J. B., J. E. Smith, Jr., R. B. Dean, E. Grossman, and O. L. Grant: "Natural Freezing for Dewatering of Aluminum Hydroxide Sludges," *J. Am. Water Works Assoc.,* 62(12):787–794, December 1970.
9. Jenkins, T. F., and A. J. Palazzo: *Wastewater Treatment by a Prototype Slow Rate Land Treatment System,* CRREL Rep. 81-14, Cold Regions Res. & Eng. Lab., Hanover, N.H., August 1981.
10. Kuter, G. A., H. A. J. Hoitink, and L. A. Rossman: "Effects of Aeration and Temperature on Composting of Municipal Sludge In a Full-Scale Vessel System," *J. Water Pollution Control Fed.,* 57(4):309–315, April 1985.
11. Lang, L. E., J. T. Bandy, and E. D. Smith: *Procedures for Evaluating and Improving Water Treatment Plant Processes at Fixed Army Facilities,* Report of the U.S. Army Construction Engineering Research Laboratory, Champaign, Ill., 1985.
12. Loehr, R. C., J. H. Martin, E. F. Neuhauser, and M. R. Malecki: *Waste Management Using Earthworms—Engineering and Scientific Relationships,* National Science Foundation ISP-8016764, Cornell University, Ithaca, N.Y., March 1984.
13. Loehr, R. C., and J. Ryan: *Land Treatment Practices in the Petroleum Industry,* American Petroleum Institute, Washington, D.C., June 1983.
14. Metcalf and Eddy Inc.: *Wastewater Engineering: Treatment, Disposal, Reuse,* 2d ed., McGraw-Hill, N.Y., 1979.
15. Miller, F. C., and M. S. Finstein: "Materials Balance in the Composting of Wastewater Sludge as Affected by Process Control Strategy," *J. Water Pollution Control Fed.,* 57(2):122–127, February 1985.
16. Pietz, R. I., J. R. Peterson, J. E. Prater, and D. R. Zenz: "Metal Concentrations in Earthworms from Sewage Sludge Amended Soils at a Strip Mine Reclamation Site," *J. Environ. Qual.,* 13(4):651–654, Apr. 1984.

17. *Process Design Manual: Municipal Sludge Landfills,* Environmental Protection Agency, EPA 625/1-78-010 (available as PB-279-675 from National Technical Information Service, Springfield, Va.), October 1978.
18. *Process Design Manual: Land Treatment of Municipal Wastewater,* EPA 625/1-81-013, Environmental Protection Agency, Cincinnati, October 1981.
19. *Process Design Manual: Sludge Treatment and Disposal,* EPA 625/1-79-011, Environmental Protection Agency, Cincinnati, September 1979.
20. *Process Design Manual: Dewatering Municipal Wastewater Sludges,* EPA 625/1-82-014, Environmental Protection Agency, Cincinnati, October 1982.
21. *Process Design Manual: Land Application of Municipal Sludge,* EPA 625/1-83-016, Environmental Protection Agency, Cincinnati, 1983.
22. Reed, S. C., J. Bouzoun, and W. S. Medding: "A Rational Method for Sludge Dewatering via Freezing," in *Comptes rendus, 7e symposium sur le traitement des eaux usées,* Montreal, Nov. 20–21, 1984, pp. 109–117.
23. Reed, S. C., and R. W. Crites: *Handbook of Land Treatment Systems for Industrial and Municipal Wastes,* Noyes Publications, Park Ridge, N.J., 1984.
24. Rush, R. J., and A. R. Stickney: *Natural Freeze-Thaw Sludge Conditioning and Dewatering,* Rep. EPS 4-WP-79-1, Environment Canada, Ottawa, January 1979.
25. Schleppenbach, F. X.: *Water Filtration at Duluth, Minnesota,* EPA 600/2-84-083 (available from National Technical Information Service, Springfield, Va, PB 84-177 807), Environmental Protection Agency, Cincinnati, August 1983.
26. Schneiter, R. W., E. J. Middlebrooks, R. S. Sletten, and S. C. Reed: *Accumulation, Characterization and Stabilization of Sludges from Cold Regions Lagoons,* CRREL Special Report 84-8, U.S. Army Cold Regions Res. & Eng. Lab., Hanover, N.H., April 1984.
27. Sittig, M.: *Landfill Disposal of Hazardous Wastes and Sludges,* Noyes Data Corp., Park Ridge, N.J., 1979.
28. Sommers, L. E., C. F. Parker, and G. J. Meyers: *Volatilization, Plant Uptake and Mineralization of Nitrogen in Soils Treated with Sewage Sludge,* Tech. Rep. 133, Purdue University Water Resources Research Center, West Lafayette, Ind., 1981.
29. Sopper, W. E., and S. N. Kerr (eds): *Utilization of Municipal Sewage Effluent and Sludge on Forest and Disturbed Land,* Pennsylvania State University Press, University Park, 1979.
30. Stotzky, G: "Microbial Respiration," in *Methods of Soil Analysis—Part 2. Chemical and Microbial Properties,* American Society of Agronomy, Madison, Wis., 1965, pp. 1550–1572.
31. US Army: *Technical Manual—Hazardous Waste Land Disposal and Land Treatment Facilities,* TM 5-814-7, Huntsville Division, U.S. Army Corps of Engineers, Huntsville, Ala., August 1984.
32. U.S. Department of Agriculture and Environmental Protection Agency: *Manual for Composting Sewage Sludge by the Beltsville Aerated Pile Method,* EPA 600/8-80-022, EPA Municipal Environmental Research Laboratory, Cincinnati, May 1980.
33. U.S. Environmental Protection Agency: *Sludge Composting and Improved Incinerator Performance,* Technology Transfer Seminar Report, Center for Environmental Research Information, Cincinnati, July 1984.
34. U.S. Environmental Protection Agency: *Composting Processes to Stabilize and Disinfect Municipal Sewage Sludge,* EPA 430/9-81-011, Office of Water Program Operations, Washington, June 1981.
35. U.S. Environmental Protection Agency, U.S. Department of Agriculture, Food & Drug Administration: *Land Application of Municipal Sewage Sludge for the Production of Fruits and Vegetables—a Statement of Federal Policy and Guidance,* EPA Office of Municipal Pollution Control, Washington, 1981.
36. Whiting, D. M.: *Use of Climatic Data in Design of Soil Treatment Systems,* EPA 660/2-75-018, Environmental Protection Agency Corvallis Environmental Research Laboratory, Corvallis, Oreg., September 1975.

Appendix

TABLE A.1 Metric Conversion Factors (SI Units to U.S. Customary System Units)

Multiply the SI unit		by	To obtain the USCS unit	
Name	Symbol		Symbol	Name
		Area		
hectare (10,000 m^2)	ha	2.4711	acre	acre
square centimeter	cm^2	0.1550	in^2	square inch
square kilometer	km^2	0.3861	mi^2	square mile
square kilometer	km^2	247.1054	acre	acre
square meter	m^2	10.7639	ft^2	square foot
square meter	m^2	1.1960	yd^2	square yard
		Energy		
kilojoule	kJ	0.9478	Btu	British thermal unit
joule	J	2.7778×10^{-7}	kWh	kilowatt-hour
megajoule	MJ	0.3725	hp · h	horsepower-hour
		Flow Rate		
cubic meters per day	m^3/day	264.1720	gal/day	gallons per day
cubic meters per day	m^3/day	2.6417×10^{-4}	million gal/day	millions gallons per day
cubic meters per second	m^3/s	35.3157	ft^3/s	cubic feet per second
cubic meters per second	m^3/s	22.8245	million gal/day	million gallons per day
cubic meters per second	m^3/s	15.8503	gal/min	gallons per minute
liters per second	L/s	22.8245	gal/day	gallons per day

ABLE A.1 Metric Conversion Factors (SI Units to U.S. Customary System Units) (*Continued*)

Multiply the SI unit		by	To obtain the USCS unit	
Name	Symbol		Symbol	Name
		Length		
entimeter	cm	0.3937	in	inch
ilometer	km	0.6214	mi	mile
ieter	m	39.3701	in	inch
ieter	m	3.2808	ft	foot
ieter	m	1.0936	yd	yard
iillimeter	mm	0.03937	in	inch
		Mass		
ram	g	0.0353	oz	ounce
ram	g	0.0022	lb	pound
ilogram	kg	2.2046	lb	pound
ietric ton (10^3 kg)	metric ton	1.1023	ton	ton (short: 2000 lb)
etric ton	metric ton	0.9842	ton	ton (long: 2240 lb)
		Power		
ilowatt	kW	0.9478	Btu/s	British thermal units per second
ilowatt	kW	1.3410	hp	horsepower
		Pressure		
ascal	Pa (N/m^2)	1.4505×10^{-4}	lb/in^2	pounds per square inch
		Temperature		
egree Celsius	°C	1.8 (°C) + 32	°F	degree Fahrenheit
elvin	K	1.8 (K) − 459.67	°F	degree Fahrenheit
		Velocity		
ilometers per second	km/s	2.2369	mi/h	miles per hour
ieters per second	m/s	3.2808	ft/s	feet per second
		Volume		
ibic centimeter	cm^3	0.0610	in^3	cubic inch
ibic meter	m^3	35.3147	ft^3	cubic foot
ibic meter	m^3	1.3079	yd^3	cubic yard
ibic meter	m^3	264.1720	gal	gallon
ibic meter	m^3	8.1071×10^{-4}	acre-ft	acre foot
ter	L	0.2642	gal	gallon
ter	L	0.0353	ft^3	cubic foot
ter	L	33.8150	oz	ounce (U.S. fluid)
egaliter	ML	0.2642	10^6 gal	million gallons

TABLE A.2 Conversion Factors for Commonly Used Design Parameters

Multiply the SI unit		by	To obtain the USCS	
Parameter	Symbol		Symbol	Parameter
cubic meters per second	m^3/s	22.727	10^6 gal/day	million gallons per day
cubic meters per day	m^3/day	264.1720	gal/day	gallons per day
kilograms per hectare	kg/ha	0.8922	lb/acre	pounds per acre
metric tons per hectare	metric tons/ ha	0.4461	ton/acre	ton (short) per acre
cubic meters per hectare per day	m^3/(ha · day)	106.9064	gal/(acre · day)	gallons per acre per day
kilograms per square meter per day	kg/(m^2 · day)	0.2048	lb/(ft^2 · day)	pounds per square foot per day
cubic meters (solids)/ 10^3 cubic meters (liquid)	$m^3/10^3\ m^3$	133.681	$ft^3/10^6$ gal	cubic feet per million gallons
m^3 (liquid)/m^2 (area)	m^3/m^2	24.5424	gal/ft^2	gallons per square foot
g (solids)/m^3 (liquid)	g/m^3	8.3454	lb/10^6 gal	pounds per million gallons
m^3 (air)/[m^3 (liquid) · min]	m^3/(m^3 · min)	1000.0	$ft^3/10^3$ min	ft^3 of air per minute per 1000 cubic feet
kW/$10^3\ m^3$ (tank volume)	kW/$10^3\ m^3$	0.0380	hp/$10^3\ ft^3$	horsepower per 1000 cubic feet
kilograms per cubic meter	kg/m^3	62.4280	lb/$10^3\ ft^3$	pounds per 1000 ft^3
cubic meters per capita	m^3 per capita	35.3147	ft^3 per capita	cubic feet per capita
bushels per hectare	bu/ha	0.4047	bu/acre	bushels per acre

TABLE A.3 Physical Properties of Water

Temperature, °C	Density, kg/m^3	Dynamic viscosity, (N · s)/m^2	Kinematic viscosity, m^2/s
0	999.8	1.781	1.785
5	1000.0	1.518	1.519
10	999.7	1.307	1.306
15	999.1	1.139	1.139
20	998.2	1.002	1.003
25	997.0	0.890	0.893
30	995.7	0.798	0.800
40	992.2	0.653	0.658
50	988.0	0.547	0.553
60	983.2	0.466	0.474
70	977.8	0.404	0.413
80	971.8	0.354	0.364
90	965.3	0.315	0.326
100	958.4	0.282	0.294

TABLE A.4 Oxygen Solubility in Fresh Water

Temperature, °C	Oxygen solubility,* mg/L
0	14.62
1	14.23
2	13.84
3	13.48
4	13.13
5	12.80
6	12.48
7	12.17
8	11.87
9	11.59
10	11.33
11	11.08
12	10.83
13	10.60
14	10.37
15	10.15
16	9.95
17	9.74
18	9.54
19	9.35
20	9.17
21	8.99
22	8.83
23	8.68
24	8.53
25	8.38
26	8.22
27	8.07
28	7.92
29	7.77
30	7.63

*Concentrations of saturated solutions of oxygen in fresh water exposed to dry air containing 20.90 percent oxygen under a total pressure of 760 mmHg.

Index

About the Authors

SHERWOOD C. REED has over 20 years of environmental research experience at the United States Army Cold Regions Research and Engineering Laboratory in Hanover, N.H. His focus has been on land treatment of wastewater and sludges, aquaculture and other innovative treatment concepts. For 12 years, he served as a special advisor to the U.S. EPA and has assisted the EPA in preparation of several design manuals. Reed has authored or contributed to 10 books and more than 70 technical articles on wastewater and sludge management.

DR. E. JOE MIDDLEBROOKS is a teacher, college administrator, researcher, author of numerous books and articles, and a recognized authority on lagoons and aquatic systems. He is currently provost and vice president for academic affairs at Tennessee Technological University.

RONALD W. CRITES is a consulting engineer with Nolte and Associates and is internationally recognized for his work on land application of wastes. He has also written several books and numerous journal articles.